Solar Cells Development and Fabrication

This book covers the basic scientific background of solar cells, their principles, working, growth, operating parameters, commercialization status, manufacturing challenges, and future scope of solar cells. Topics covered range from history and developments of solar cell generation to market growth and different applications of solar cells including in-depth knowledge about Si, PSCs, and next generation multilayer bandgap based solar cells and their fabrication techniques with advanced methodology.

Key features:

- Explains solar cells and their growth at different stages.
- Discusses challenges in the fabrication/commercialization of solar cells at the lab and industry levels.
- Combines fundamental, experimental, and theoretical knowledge with industrial needs and engineering design methods.
- Covers the new generation of perovskite solar cells and their synthesis techniques.
- Explores multilayer graded bandgap solar cells and their importance in existing solar technology.

This book is specifically designed for graduate students and researchers in solar energy technology, cell device, and materials science.

AF763671

Emerging Materials and Technologies

Series Editor: Boris I. Kharissov

The *Emerging Materials and Technologies* series is devoted to highlighting publications centered on emerging advanced materials and novel technologies. Attention is paid to those newly discovered or applied materials with potential to solve pressing societal problems and improve quality of life, corresponding to environmental protection, medicine, communications, energy, transportation, advanced manufacturing, and related areas.

The series takes into account that, under present strong demands for energy, material, and cost savings, as well as heavy contamination problems and worldwide pandemic conditions, the area of emerging materials and related scalable technologies is a highly interdisciplinary field, with the need for researchers, professionals, and academics across the spectrum of engineering and technological disciplines. The main objective of this book series is to attract more attention to these materials and technologies and invite conversation among the international R&D community.

Metal Organic Framework Derived Materials: Design Strategies and Applications
Gomathi Nageswaran, Varsha M V, Arun Kumar Rajasekaran, and M Shashank Rao

Hydrogen Production, Storage, and Utilization: Technologies and Applications
Abbas Tcharkhtchi, Hamidreza Vanaei, Albert Lucas, and Sedigheh Farzaneh

MXenes for Energy Storage Applications: Emerging Characteristics, Compositions, and Synthesis Methods
Muhammad Rafique, M. Bilal Tahir, and Saira Anwar

Multi-scale and Multifunctional Coatings and Interfaces for Tribological Contacts
Ajit Behera, Kuldeep K Saxena, Dipen Kumar Rajak, and Shankar Sehgal

Nanotechnology in Green Energy Generation
Ahmed Thabet Mohamed

Thermo-Acoustics of Nanofluids and Transfer Processes
Shriram S. Sonawane and Manjakuppam Malika

Hybrid Nanofluids: Heat and Mass Transfer Processes
Shriram S. Sonawane and Manjakuppam Malika

Applications of Hybrid Nanofluids in Science and Engineering
Edited by A. K. Pandey, H. Upreti, O. D. Makinde, and A. J. Chamkha

Solar Cells Development and Fabrication
Edited by Shivani Dhall, Kapil Sood, and Vinay Gupta

For more information about this series, please visit: www.routledge.com/Emerging-Materials-and-Technologies/book-series/CRCEMT

Solar Cells Development and Fabrication

Edited by

Shivani Dhall, Kapil Sood, and Vinay Gupta

CRC Press
Taylor & Francis Group
Boca Raton London New York

CRC Press is an imprint of the
Taylor & Francis Group, an **Informa** business

Designed cover image: Shutterstock

First edition published 2025
by CRC Press
2385 NW Executive Center Drive, Suite 320, Boca Raton FL 33431

and by CRC Press
4 Park Square, Milton Park, Abingdon, Oxon, OX14 4RN

CRC Press is an imprint of Taylor & Francis Group, LLC

ISBN: 978-1-032-79659-8 (hbk)
ISBN: 978-1-032-81408-7 (pbk)
ISBN: 978-1-003-49970-1 (ebk)

DOI: 10.1201/9781003499701

Typeset in Times
by codeMantra

Contents

Acknowledgments

I, Dr. Shivani Dhall, would like to acknowledge the extraordinary efforts of all the authors, Publisher CRC Press, Dr. Gagandeep Singh and Dr. Jahnavi Vaid for their enormous skills and precession in writing this book. Dr. Shivani Dhall acknowledges UGC-DAE-CSR for providing funding to support her research work (sanctioned no. CRS/2023-24/03/1206). I would like to express my deepest appreciation to Dr. P. D. Babu, Centre-Director UGC-DAE Consortium for Scientific Research, and Dr. Jyoti Prakash, Senior Scientist in Bhabha Atomic Research Centre, Mumbai, India, for their continuous encouragement for this assignment. This endeavor would not have been possible without the support of Dr. Rajesh Kumar, Principal, D.A.V. College, Jalandhar, who is a continuous source of inspiration in pursuing the research assignments. I would not be able to get this work done without the continual support and vision of Co-Editors Professor Vinay Gupta and Dr. Kapil Sood. Dr. Vinay Gupta would like to thank Khalifa University for kind support. I wish to thank God, my parents, my husband, and my loving daughters Shritika and Harshika for their continuing love and support.

Preface

Nowadays, the significance of connecting the power from the sun to ensure our energy demands without disturbing the planet is extensively considerable.

Developments in human civilization have revolved around the consumption of energy. Historically, control over energy has been directly related to the centralization of power. The natural abundance of fossil fuels has helped sovereign nations in gaining better control over sociological developments. The absence of such natural resources has forced the remaining societies to greatly depend on the lucky ones that have them. In some cases, the developments have slowed to a crawl, and even basic needs have not been fulfilled. Therefore, materials scientists and engineers have taken up the challenge to reduce our dependence on fossil fuel resources. The invention of methods to harness solar energy has given new hope to the nations with limited or no fossil fuel reserves. In the last few decades, the idea of technologies based on solar cells has been well established. New techniques of materials synthesis and their integration in novel engineering designs have helped the industry produce solar cells with high energy efficiency. The use of high-purity silicon has dominated the solar cell industry for several decades. It appears that the efficiency of silicon-based solar cells has achieved maximum potential and no further improvements are expected. The high cost of extra pure silicon and saturation in energy efficiency has motivated scientists to look for new materials and technologies that are efficient and relatively cost effective.

Journals are flooded with research articles with claims about materials and technologies that might be able to produce highly efficient solar cells. It is, therefore, important to summarize the literature in order to draw meaningful conclusions based on the latest research findings. Since silicon-based solar cells are immensely popular, various books are available on the subject. It was, therefore, decided to look at other types of solar cells, which are either in the development phase or are likely competitors of silicon solar cells.

The purpose of our efforts has been to collect articles from various experts in the field with this view in mind. This book is composed of six chapters, each of which was written by an expert in their field. Topics are broadly designed to cover dye-sensitized types of solar cells and their related problems, layered types of solar cells, application of lithography in solar cells, and luminescent solar and plasmonic light trapping. Graphene being the most recent discovery, its application in organic types of solar cells is covered in this book. Finally, analytical modeling and electrical circuit design, which are other important aspects of solar cell development, are also included. The final section of this book appends to the series of articles written on putative future trends in this area. This section is specifically written for advanced practitioners to help them channel their expertise in the desired direction.

We expect that readers will gain in-depth knowledge in new areas of solar cells, which are not commonly known and for which literature is scarcely available. This title has been edited for a broad readership that includes scholars and researchers from diverse backgrounds and for commercial sectors looking for innovative solar cell materials and related technologies. This could be a solitary reference book for researchers in materials science, engineering, and nanotechnology. We are confident that readers will gain invaluable information that can be utilized not only in research and development but also in the commercial manufacturing of solar cell panels.

Shivani Dhall

About the Editors

Dr. Shivani Dhall has been working as an assistant professor in the Department of Physics, D.A.V. College, Jalandhar, since July 2018. Previously, she had a National Post Doctoral Fellowship (NPDF) in the Department of Physics, Indian Institute of Technology (IIT) Delhi. During her Ph.D., she worked in Centre of Excellence in Nanoelectronics (CEN), Electrical Department, Indian Institute of Technology Bombay. She got the best Indian Nanoelectronics Users' Programme (INUP) project award from IIT Bombay in 2015. In 2018, she was awarded with Council of Scientific and Industrial Research (CSIR) Nehru and DS Kothari Fellowships. In 2019, she received a Teachers Associateship for Research Excellence (TARE) project under Department of Science and Technology (DST) to continue her research work. She is working on carbon materials, nanoparticles, and their device fabrication using lithography technique for gas sensing, photo-sensor, and interconnected applications. She is a reviewer of many Science Citation Index (SCI) journals and has 30 Science Citation Index journal publications.

Dr. Kapil Sood has been working as an assistant professor in the Department of Physics, Government Degree College Dhaliara, Teh. Dehra, Distt. Kangra, Himachal Pradesh, India. Upon receiving his Ph.D. from the School of Physics and Materials Science, Thapar University Patiala, Punjab, in 2014, he worked as a postdoctoral fellow in the Department of Chemical Engineering, Indian Institute of Technology Delhi (2014–2018). He also worked as a visiting researcher with Professor Truls Norby, Department of Chemistry, University of Oslo, Norway. He has published many papers in various journals of international repute. His main areas of research are renewable energy sources, electrochemical energy devices, solid electrolytes, and fuel cells.

Vinay Gupta is an Indian materials scientist and a former senior scientist at National Physical Laboratory of India. He is known for his work on organic solar cells, carbon nanotube arrays, and Förster resonance energy transfer, Gupta is a former Alexander von Humboldt Fellow and JSPS Fellow. He is the recipient of Thomson Reuters India citation award 2015. The Council of Scientific and Industrial Research, the apex agency of the Government of India for scientific research, awarded him the Shanti Swarup Bhatnagar Prize for Science and Technology, one of the highest Indian science awards, for his contributions to physical sciences in 2017. He is currently working in the Department of Physics, Khalifa University of Science and Technology, Abu Dhabi, UAE.

List of Contributors

Anil Astakala
Department of Transportation Energy Convergence
Korea National University of Transportation
Chungju, Republic of Korea

Swati Bishnoi
Department of Chemical Engineering
Indian Institute of Technology
Delhi, India

Ram Datt
Faculty of Science and Engineering, SPECIFIC
Swansea University
Swansea, United Kingdom

Shivani Dhall
Department of Physics
D.A.V. College
Jalandhar, Punjab, India

Mamta Goel
Department of Physics
DAV University
Jalandhar, Punjab, India

Vinay Gupta
Department of Physics
Khalifa University
Abu Dhabi, United Arab Emirates

Summit Jalota
Department of Physics
D.A.V. College
Jalandhar, Punjab, India

Naveen Kumar
Department of Physics
Government Degree College
Haripur Guler, Himachal Pradesh, India

Neetesh Kumar
Department of Mechanical and Materials Engineering
NASA CRE2DO Center
Miami, Florida

Abhinav Nag
Department of Physics
GGDSD College
Palampur, Himachal Pradesh, India

Jyoti Prakash
Materials Group
Bhabha Atomic Research Centre
Trombay, Mumbai, Maharashtra, India

Shilpa Setia
Department of Chemistry
D.A.V. College
Jalandhar, Punjab, India

Krupali Shetty
Continental Automotive Corporation
Electronic City, Bengaluru, Karnataka, India

Kapil Sood
Department of Physics
Government College Dhaliara
Dhaliara, Himachal Pradesh, India

S.K. Tripathi
Department of Physics
Center of Advanced Study in Physics, Panjab University
Chandigarh, India

Wing Chung Tsoi
SPECIFIC, Faculty of Science and Engineering
Swansea University
Swansea, United Kingdom

Barkha Tyagi
Department of Physics
University of Oxford
Oxford, United Kingdom

1 Development of Solar Cells

*Swati Bishnoi, Ram Datt, Wing Chung Tsoi, and Vinay Gupta**

1.1 INTRODUCTION

The evolution of contemporary energy demands has driven exploration into novel technologies and physical and chemical processes aimed at establishing efficient systems for energy generation, storage, transformation, and transportation. Despite the ongoing availability of global fossil fuel reserves, the detrimental social, health, and environmental repercussions of current unsustainable energy consumption patterns are increasingly evident [1]. Consequently, the imperative for large-scale adoption of alternative energy production methods to sustain and elevate our standard of living has become apparent. Continued adherence to present trends suggests an escalating need for electrical energy in future societies. The likelihood of substantial global warming due to greenhouse gas emissions within the next five decades is high. Climate change represents an imminent and formidable threat necessitating collective action. Fortunately, advancements in science and technology have yielded various sustainable alternatives for energy production, including wind, geothermal, biomass, and solar energy [2]. In response to the pressing need for sustainable energy solutions and the imperative to reduce dependence on fossil fuels, concerted global efforts have been directed toward advancing clean energy technologies. Among these, solar energy stands out as a promising and abundant resource for electricity generation. Solar cells utilize the photovoltaic effect to transform sunlight into electrical energy, representing a significant eco-friendly power source. Over the past two decades, there has been a continuous evolution in solar cell technologies aimed at enhancing power conversion efficiency (PCE) [3]. Since Alexandre-Edmond Becquerel's discovery of the photovoltaic (PV) effect in 1839, harnessing solar energy has been a focal point of scientific communities. The energy absorbed by the earth's atmosphere from the sun every hour exceeds global energy requirements for an entire year, driving intensified research efforts in recent decades to develop efficient and cost-effective solar cells. Solar cells offer numerous advantages, including their renewable nature and global availability. Unlike conventional energy sources such as coal, nuclear, oil, and gas, solar cells incur no fuel costs and require relatively low operation and maintenance expenses, ensuring indefinite power generation as long as sunlight persists. Consequently, solar cells represent a sustainable and environmentally friendly energy production method, offering the added advantage of being cost-free.

DOI: 10.1201/9781003499701-1

1.2 BACKGROUND AND IMPORTANCE OF SOLAR CELLS

Solar cells are based on the photovoltaic effect, which occurs when certain materials, known as semiconductors, absorb photons from sunlight. The absorption of photons generates electron-hole pairs, where electrons are excited from low energy to high energy levels and, after charge carrier separation, create a flow of electric current when the material is connected to an external circuit. Solar cells must capture a broad range of energy corresponding to the solar spectrum for optimal efficiency. The solar spectrum spans wavelengths from 100nm to 1mm, with peak irradiance occurring between 250 and 2500nm, predominantly within the visible light range (380–700nm). The light intensity was calculated using Air Mass (AM) conditions. While considering light intensity on earth, the defined AM is called AM1.5G (global) and AM1.5D (direct). Meanwhile, the defined AM for aerospace is AM0, and the light intensity is a bit higher than the earth's surface; for AM1.5G/D, sunlight is attenuated by passing the light through the atmosphere. The values of light intensities are 100 and 133.6 mW/cm^2 under AM1.5G and AM0, respectively. Evaluating the efficacy of photovoltaic technologies often hinges on conversion efficiency, a parameter gauging the proportion of solar energy input converted into electrical energy output. This parameter encompasses several system attributes including short-circuit current, open-circuit voltage, and fill factor, all of which are contingent upon intrinsic material properties and fabrication techniques. Figure 1.1 demonstrates the solar cell's photovoltaic parameters, defined in the paragraph below.

1.2.1 Open Circuit Voltage

The open circuit voltage (V_{OC}) signifies the maximum voltage attainable from a solar cell with its terminals unconnected or the voltage achievable from a cell when no current (open-circuit condition, $J_{SC} = 0$) flows through solar cells.

1.2.2 Short Circuit Current Density

The short circuit current density (J_{SC}) defines a current flow to the solar cells under short-circuit conditions ($V_{OC} = 0$). It characterizes the available current when both terminals of a cell are shorted.

1.2.3 Fill Factor

The fill factor (FF) quantifies the ratio of maximum power (P_{max}) to the product of J_{SC} and V_{OC}. FF relies on resistive losses within the solar cell. An ideal solar cell, devoid of any losses, would exhibit an FF of 100%, although such perfection is unattainable. Materials with a high bandgap tend to exhibit higher FF.

1.2.4 Power Conversion Efficiency

Power conversion efficiency (PCE) delineates the ratio of output to input power, denoted by η. It can be computed as the ratio of maximum power output (P_{max}) to power input (P_{in}):

$$\eta = P_{max} / P_{in} = V_{OC} * J_{SC} * \mathrm{FF} / P_{in}$$

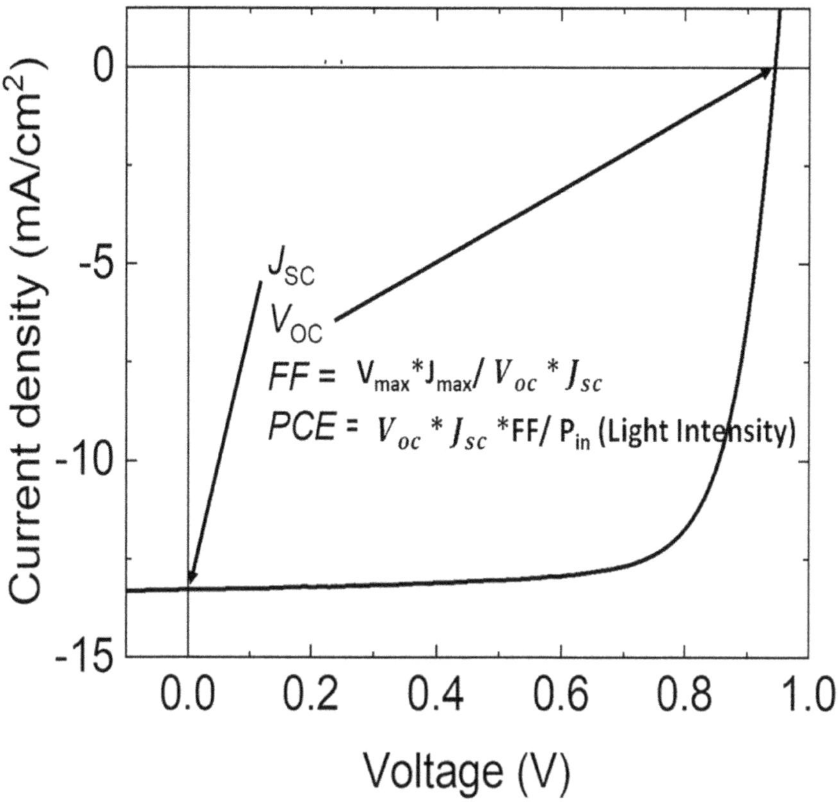

FIGURE 1.1 J-V characteristics (measured under light) of solar cells.

The cost-effectiveness and efficiency of solar cells depend on the materials used in their fabrication. Research has aimed to identify materials that balance efficiency and cost for PV solar cells. A significant challenge in photovoltaic cell development is achieving maximum efficiency while minimizing production costs. Enhancing solar cell efficiency involves addressing internal losses, primarily from optical, and electrical sources. The upper-efficiency threshold for commercially accessible technologies is defined by the Shockley–Queisser (SQ) limit, which factors in the equilibrium between photogeneration and radiative recombination [4]. Reducing electrical losses holds the greatest potential for efficiency improvement, as they are closely tied to material properties and cell structure.

1.3 SOLAR CELL GENERATION

Solar cells are categorized based on the materials employed in their fabrication. Silicon is predominantly utilized in the production of single-crystalline solar cells. It also serves as a vital component in polycrystalline silicon solar cells and amorphous silicon solar cells. Thin film solar cell technology utilizes various materials such as amorphous silicon (a-Si), cadmium sulfide (CdS), cadmium telluride (CdTe), and copper

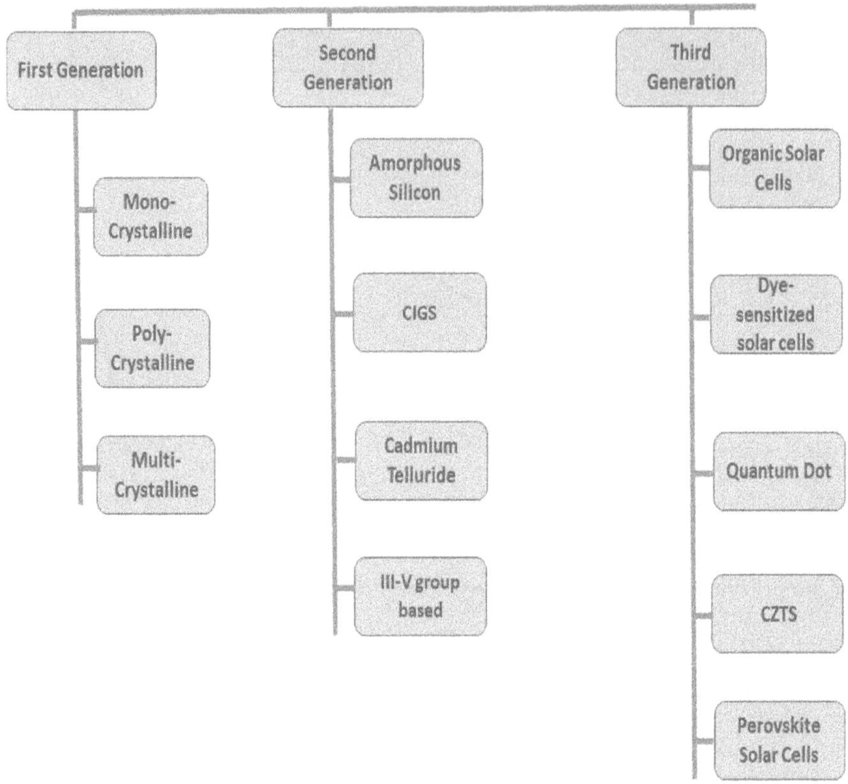

FIGURE 1.2 Hierarchical diagram illustrating the evolution of solar cell generations and sub-generations.

indium gallium di-selenide (CIGS) [5]. Third-generation solar cell technologies employ advanced principles to maximize solar energy absorption and enhance photon conversion efficiency. This category encompasses quantum dot solar cells, organic solar cells (OSCs), perovskite solar cells (PSCs), and dye-sensitized solar cells (DSSCs). Figure 1.2 shows the classification of different solar cell generations. The efficiencies of the different generation solar cells, along with their historical timeline are given in Figure 1.3. The graph illustrates the efficiency rates across various generations, providing a comprehensive overview of the state-of-the-art advancements in solar energy conversion.

1.3.1 First Generation Solar Cells

The first generation refers to solar cells predominantly comprising mono- or polycrystalline wafer-based silicon cells. Monocrystalline silicon cells involve the growth of silicon blocks from small monocrystalline seeds and subsequent cutting to produce monocrystalline silicon wafers, utilizing the Czochralski process. Monocrystalline material is preferred due to its higher efficiency compared to multi-crystalline material. Key technological challenges associated with

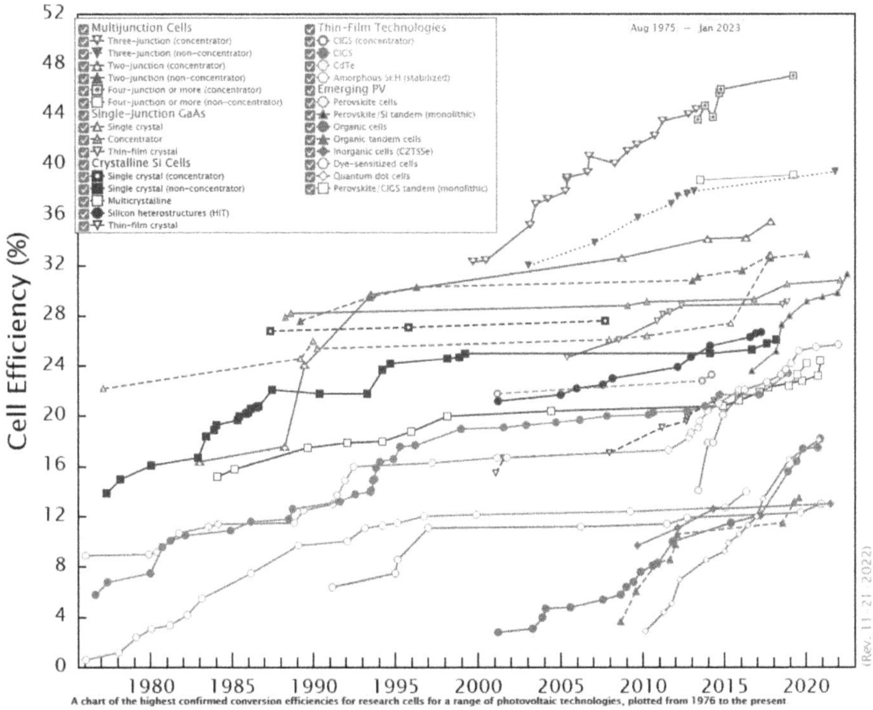

FIGURE 1.3 NREL Solar Cell Efficiency Data Curve showcasing the current efficiencies of prevalent solar cell technologies [6].

monocrystalline silicon include stringent requirements for material purity, significant material consumption during cell fabrication, manufacturing processes, and limitations on module sizes comprising these cells [7]. Multicrystalline silicon blocks are generated by melting high-purity silicon and crystallizing it in a large crucible through the directional solidification process. Unlike the Czochralski process, there is no reference crystal orientation in this method, resulting in silicon material with varied orientations. P-type silicon substrates doped with boron are primarily used as the base material for solar cells. While n-type silicon substrates are employed in the production of high-efficiency solar cells, and present extra technical hurdles, notably in achieving uniform doping throughout the silicon block when compared to p-type substrates. Manufacturing crystalline solar cells involves a sequential execution of six or more steps, including surface texturing, doping, diffusion, oxide removal, anti-reflective coating, metallization, and firing. Afterward, cell efficiency and other parameters are assessed under standard test conditions. The efficiency of photovoltaic cells is heavily influenced by the quality of materials used in their fabrication.

1.3.1.1 Mono-Crystalline Silicon Solar Cells

Mono-crystalline Si solar cells are made from a single crystal of silicon, which is grown from a seed crystal in a controlled environment. This manufacturing process results in a high-purity silicon crystal structure with uniform electrical properties, leading to higher efficiency compared to other types of solar cells. Mono-crystalline Si solar cells typically have a higher PCE and a smaller footprint, making them suitable for applications where space is limited, such as residential rooftops.

1.3.1.2 Poly-Crystalline Silicon Solar Cells

Poly-crystalline Si solar cells are similar to semi-crystalline Si solar cells in that they are made from silicon wafers containing multiple crystal grains. However, poly-crystalline silicon wafers are produced using a different manufacturing process called the casting method, which involves pouring molten silicon into a mold and allowing it to solidify. This results in a material with larger crystal grains compared to semi-crystalline silicon but with more defects and grain boundaries. As a result, poly-crystalline silicon solar cells typically have lower efficiency and a lesser fabrication cost compared to mono-crystalline ones. Cell efficiency is constrained by factors such as volume defects. Production breakthroughs include the implementation of an aluminum back surface field (Al-BSF) to mitigate the recombination rate on the back surface or the advancement of Passivated Emitter and Rear Cell (PERC) technology to further diminish the recombination rate on the back surface [8]. The initial iteration of solar cells revolves around those constructed using crystalline silicon wafers. These cells are crafted from crystalline silicon, a process which entails the most expensive technique for obtaining pure silicon crystal.

1.3.2 Second Generation Solar Cells

Second-generation solar cells, particularly thin-film solar cell technology, have garnered significant attention due to their potential to reduce the cost of solar cells compared to traditional crystalline Si wafer technology. The conventional Si crystal wafer technology relies heavily on high-purity crystalline Si, which is obtained through a complex and expensive process. However, thin-film technology offers a promising alternative by allowing for the deposition of thin layers of Si, typically around 1 μm thick, thus minimizing the amount of silicon required.

One of the pioneering advancements in thin-film technology was the development of amorphous silicon (a-Si) deposited thin films by R. Chittick and his team [9]. Amorphous silicon is a non-crystalline form of silicon with unique electronic properties that make it suitable for solar cell applications. Moreover, CIGS, CdTe, and solar cells based on III-V group materials also come under the category of third generations.

1.3.2.1 Amorphous Silicon Solar Cells

The amorphous silicon (a-Si: H) is dominated by thin film-based solar cells. The solar cells are made of either p-i-n or n-i-p device structure, wherein p-layer and n-layer serve to establish an internal electric field (i-layer) comprising amorphous silicon. Given its high absorption capacity, the i-layer typically ranges in thickness from 0.2 to 0.5 μm. The doping capacity of amorphous silicon is notably restricted. In traditional PN junction solar cells, minority carriers generated from photo-absorption diffuse throughout the P and N regions across the entire junction length, benefiting from significantly large diffusion lengths, typically around 200 µm. However, in amorphous silicon solar cells, the diffusion lengths for minority charge carriers is approximately 0.1 µm, rendering it impractical to solely rely on diffusion for collecting photogenerated charge carriers [10]. To address this challenge, p-i-n diodes are consistently employed in a-Si:H solar cells. Additionally, thin-film technology enables the manufacture of homo-junction, heterojunction, and multi-junction solar cells, broadening the range of design options available for optimizing performance and efficiency. However, compared to crystalline materials, amorphous material's structural homogeneity is relatively low, potentially leading to increased electron-hole recombination over long distances. To mitigate this, a careful balance is struck in the i-layer thickness to ensure optimal absorption without compromising efficiency, often achieved through a multilayer structured stack design. One notable limitation of photovoltaic cells is the Staebler-Wronski effect (SWE), wherein performance experiences a pronounced but temporary decline after exposure to light. This effect arises from light-induced structural alterations in certain unsaturated silicon atoms within the material [10]. Manufacture of amorphous silicon photovoltaic cells primarily relies on plasma-enhanced chemical vapor deposition (PECVD), facilitating the production of silicon thin films. Flexible and cost-effective substrates such as stainless steel or plastic materials can be employed in larger sizes. While the evaporation process is slow and the incorporation of a high-quality conductive glass layer proves costly, the overall manufacturing cost remains only marginally lower than that of crystalline solar cells.

1.3.2.2 Copper Indium Gallium Di-Selenide (CIGS) Solar Cell

Since its inception, copper indium diselenide ($CuInSe_2$) thin-film technology has stood out for its promising attributes in solar cell applications due to its favorable electronic and optical characteristics. Subsequent research revealed that by incorporating gallium (Ga) in place of indium (In), the bandgap of the material could be elevated from approximately 1.04 electron volts (eV) in copper indium diselenide (CIS) films to around 1.68 eV in copper gallium diselenide (CGS) films. Notably, optimal device performance has been achieved through partial substitution of Ga for In, resulting in a notable enhancement in overall efficiency and the attainment of an optimal bandgap. These solar cells are commonly referred to as copper indium gallium diselenide [$Cu(In_xGa_{1-x})Se_2$] or CIGS cells, and they exhibit a high absorption coefficient, necessitating a very thin layer of material [11]. Research has shown that CIGS solar films achieve high efficiencies when deposited with heavy alkali elements such as rubidium and cesium instead of sodium and potassium, leading to improved experimental outcomes [12].

During the manufacturing process, depositing CIGS films onto a substrate is commonly executed under vacuum conditions, employing either evaporative or sputtering techniques. Sequential deposition of copper, gallium, and indium, followed by annealing with selenide vapor, culminates in the formation of the final CIGS structure. Alternatively, deposition without vacuum is feasible using nanoparticles or electroplating, although further refinement of these methods is required to ensure economic viability on a large scale. Emerging approaches resemble printing technologies more closely than conventional silicon solar-cell fabrication. In one such method, a printer dispenses droplets of semiconducting ink onto an aluminum foil. Subsequent printing stages apply additional layers and the front contact atop the initial layer, with the foil subsequently being divided into sheets.

CIGS solar cells offer the advantage of manufacturability on flexible substrates, rendering them suitable for applications where conventional crystalline photovoltaics and other rigid products are impractical. For instance, flexible CIGS solar cells afford architects greater flexibility in styling and design choices. Moreover, CIGS solar cells are significantly lighter than silicon cells and can be manufactured without glass, enhancing their shatter resistance. These attributes enable integration into various vehicles such as tractor-trailers, airplanes, and automobiles, as their low profile reduces air resistance without significantly adding to the overall weight.

1.3.2.3 Cadmium Telluride Solar Cells

Cadmium telluride (CdTe) solar cells feature thin-film layers of cadmium telluride semiconductor material, facilitating the conversion of absorbed sunlight into electricity. In these solar cells, one electrode is typically composed of copper-doped carbon paste, while the other electrode is crafted from tin oxide or cadmium-based stannous oxide. Positioned between these electrodes is cadmium sulfide. CdTe photovoltaic solar cells rank as the second most abundant solar cell technology worldwide, following crystalline silicon-based solar cells. CdTe thin-film PV solar cells offer swift assembly and serve as a cost-effective alternative to conventional silicon-based PV technologies. However, the development of a stable, low-resistivity back contact for CdTe solar cells remains a challenge. Employing rapid thermal processing (RTP) to activate the back contact has resulted in a remarkable enhancement in crucial device performance metrics.

CdTe-based solar cells are categorized as thin-film technology due to their active layers being only a few microns thick, approximately one-tenth the diameter of a human hair. A schematic representation of a typical CdTe solar cell is provided in Figure 1.4. Transparent conducting oxide (TCO) layers, such as SnO_2 or Cd_2SnO_4, exhibit transparency to visible light and high conductivity to facilitate efficient current transport. Intermediate layers, such as CdS, play a dual role in facilitating growth and enhancing the electrical properties between the TCO and CdTe layers. The CdTe film serves as the primary photoconversion layer, absorbing a significant portion of visible light within the initial micron of the material. Collectively, the CdTe, intermediate, and TCO layers establish an electric field that converts absorbed light in the CdTe layer into current and voltage. Metal is applied to the back to establish electrical contacts. During production, all these layers are deposited onto incoming glass and processed into complete solar panels within a few hours.

1.3.2.4 III-V Group Based Solar Cells

III-V semiconductors are predicated on the utilization of group III or Boron group and group V or Nitrogen group elements. III-V semiconductor compounds represent a significant photovoltaic technology known for their high efficiency, albeit at a relatively higher cost. The cost is primarily attributed to the complex synthesis and device fabrication processes, compounded by the availability of relatively rare

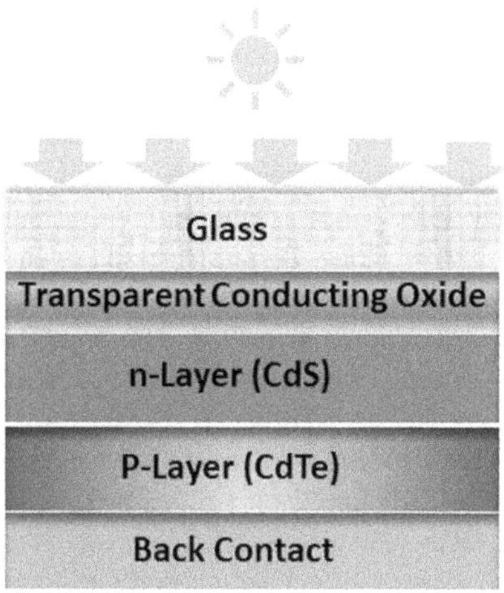

FIGURE 1.4 Schematic of a CdTe solar cell.

elements such as Indium and Gallium. Despite these challenges, III-V solar cells offer distinct advantages, chiefly owing to their unique material properties, making them the most efficient photovoltaic materials currently available. A significant factor contributing to the reduced efficiency of solar cells is energy loss resulting from the mismatch between the energy of incident photons and the energy band gap of the material utilized. When the energy content of incident photons falls below the band gap energy of the material, these photons remain unabsorbed. Conversely, if incident photons possess greater energy than the material's band gap, excess energy is lost. To mitigate these losses and enhance efficiency, the adoption of multiple layers of photovoltaic materials with a range of band gap energies is proposed, allowing for more efficient absorption of photon energy across the solar spectrum incident on the cell. Multi-junction solar cells, comprising more than one junction, are particularly favored for achieving higher efficiency.

The flexibility in combining various materials, ranging from binary to quaternary compounds, allows for precise band-gap engineering, a pivotal factor in enhancing solar cell efficiency. Additionally, many of these compounds exhibit strong interaction with light, characterized by direct band gaps and high absorption coefficients, resulting in efficient light emission. Consequently, III-V semiconductors emerge as a class of materials well-suited for optoelectronic applications, particularly in high-efficiency solar cells employing single junction concepts. Furthermore, their adaptability in tailoring band gaps to different spectra, whether global, direct concentrated, or space spectra, underscores their versatility across diverse applications. Moreover, their inherent band-gap engineering properties facilitate the exploration of new design concepts, thereby driving fundamental research and innovation in the field. Historically, III-V solar cell development has been primarily steered by niche applications necessitating high efficiency, such as space missions where lightweight, reliable energy sources are imperative. Presently, efforts are underway to extend their application to terrestrial settings, facilitated by cost-reducing solar concentrator technologies, thereby broadening their utility beyond specialized contexts.

Gallium-Arsenide (GaAs), an efficient III-V group compound with a band gap near 1.42 eV, stands out as an ideal material for highly efficient thin-film solar cells. Steiner et al. [13] demonstrated the fabrication of high-quality GaAs solar cells exhibiting beneficial back reflectance of photon energy and an impressive photon conversion efficiency of 27.8%, alongside an increased open-circuit voltage of 1.1 V. GaAs solar cells have multiple advantages, including higher photoelectric conversion efficiency, direct-gap semiconductor materials. The band-gap tailoring by controlling the composition and doping of material, superior radiation resistance.

InGaP, another semiconductor comprising indium, gallium, and phosphorus, possessing a lattice structure similar to that of GaAs, was shown to yield promising results. An InGaP solar cell based on single-junction, with a band gap energy of 1.81 eV achieved a solar conversion efficiency of 20.8% [14]. Given that a single-junction solar cell cannot effectively absorb all incident sunlight photons, multi-junction solar cells incorporating two or three junctions have been developed and demonstrated to achieve higher efficiency. The three junction cell, comprising InGaP, GaAs, and CIGS, boasts an impressive efficiency of 29.3%, with an open-circuit voltage of 2.97 V, short-circuit current density of 12.41 mA/cm^2 and a fill factor of 0.80 [15]. Led by Japan's National Institute of Advanced Industrial Science and Technology (AIST), the researchers employed a mechanical stacked design, connecting the cells via a modified smart stack with palladium (Pd) nanoparticles and adhesive. Enhancements to the bottom cell's surface were achieved through wet etching, focusing on minimizing surface roughness and thin TCO layer thickness. This meticulous approach resulted in a significant improvement in performance, surpassing previous records and marking a milestone in multijunction solar cell technology.

1.3.3 Third Generation Solar Cells

Third-generation solar cells are solution-processable with excellent potential for low-cost large-scale solar electricity generation. This generation of solar technologies offers the advantage of a tunable bandgap, enhancing their efficacy in light harvesting applications. The innovation is widely regarded as a potential paradigm shift in solar cells.

1.3.3.1 Dye Sensitised Solar Cells

DSSCs utilize cost-effective materials, including natural dyes, and offer a simpler fabrication process compared to Si solar cells, thereby presenting a strong alternative in the realm of solar cell technology. The cells can be manufactured using low-cost techniques like inkjet or screen printing and roll-to-roll processing, enabling the production of large-area devices on flexible substrates [16]. The light absorption in DSSCs is facilitated by dye molecules, with conjugated polymers and organic semiconductors demonstrating success in similar applications. The core structure of a DSSC (Figure 1.5) comprises a highly porous, nanocrystalline layer of titanium dioxide (TiO_2) acting as an electron conductor, in contact with an electrolyte solution containing organic dyes that absorb light. Charge transfer at the interface facilitates

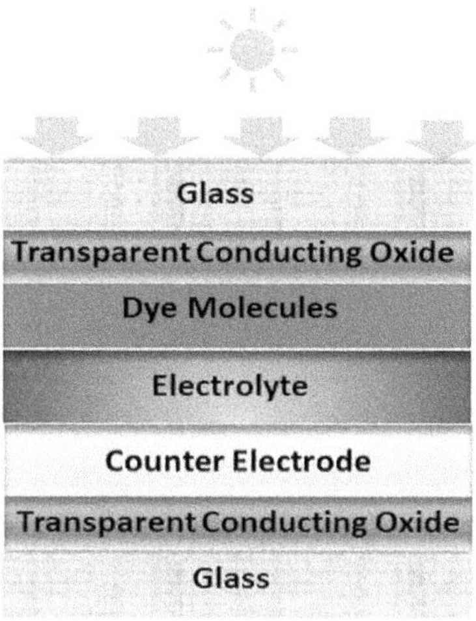

FIGURE 1.5 Schematic representation of DSSCs.

the transport of holes within the electrolyte, contributing to the overall efficiency of the cell. DSSCs have demonstrated a PCE of around 14%, under 1 sun [17].

Despite their promise, challenges such as low PCE and cell stability hinder the widespread adoption of DSSCs. Various strategies are being explored to address these challenges, including optimizing the redox pair and dye absorbance, modifying the working electrode, and developing a suitable counter electrode. Additionally, efforts are focused on reducing material costs while enhancing efficiency through techniques like co-sensitization, where multiple sensitizing dye agents are combined to utilize a broader spectrum of incident light. Metal complexes based on heavy metals such as ruthenium (Ru), osmium (Os), and iridium (Ir) are commonly used in inorganic dyes [18]. Organic dyes offer advantages in terms of cost-effectiveness, ease of preparation, and environmental friendliness. Natural dyes, including indole derivatives and zinc porphyrins, have also shown promise in enhancing DSSC efficiency [19].

1.3.3.2 Quantum Dots Solar Cells

Quantum Dot Solar Cells (QDSCs) harness tiny semiconductor particles known as quantum dots (QDs), typically a few nanometers in size, to absorb incident sunlight photons and induce the photovoltaic effect [20]. These quantum dots exhibit optical and electronic properties distinct from bulk materials, offering tunability in emitted light frequency through adjustments in size, shape, and material composition. Synthesis of quantum dots employs various techniques such as Chemical ablation, electrochemical carbonization, laser ablation, microwave irradiation, and hydrothermal/solvothermal treatment. QDSCs evolved from DSSCs, have garnered substantial research interest worldwide, attracting physicists, chemists, and engineers. These interdisciplinary efforts aim to develop novel materials, investigate photophysical processes, produce light-harvesting materials, and design unique device architectures for QDSCs, with potential breakthroughs enhanced through synergistic collaboration. Advantages of QDSCs include enhanced charge creation, separation, and

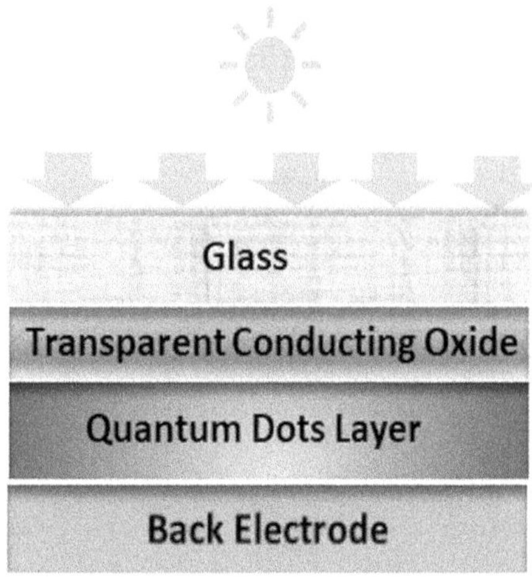

FIGURE 1.6 The working of QDSC and the mechanism of charge transfer in QDSC under illumination.

extraction within the same material, positioning them as promising nominees for high-performance solar cells. Typically, charge creation occurs within quantum dots, with subsequent rapid transportation into separate mediums, minimizing electron-hole recombination. The inception of QDSCs dates back to 2010 when Luther et al. [21] pioneered a bilayer of PbS/ZnO QDs, exhibiting good stability and achieving an efficiency of 2.94%. Subsequent research endeavors have explored various quantum dot materials to enhance QDSC efficiency. For instance, Hao et al. [22] achieved a notable PCE of 16.6% of $Cs_{1-x}FA_xPbI_3$ perovskite QD materials-based QDSCs. Components of QDSCs include Transparent conducting oxides (TCO), wide-bandgap mesoporous semiconducting materials for the photoanode, an electrolyte with a redox couple, a counter electrode (CE), and quantum dots as sensitizers. The schematic architecture of a QDSC is depicted in Figure 1.6 illustrating its layered structure.

On light exposure, in QDs there occurs the creation of electron-hole pairs QD* (e^-/h^+), a process known as charge separation. Followed by an electron injection process in which the newly created electrons and holes undergo transfer to their respective terminals. Specifically, electrons migrate into the semiconductor metal oxide, serving as an electron acceptor, while the holes are directed toward either the electrolyte or CE. The CE usually comprises either metal or semiconductor electrodes known for their enhanced catalytic activity towards the redox couple, such as Au or Pt, which efficiently accept holes from the QDs. Simultaneously, electrons within the metal oxide are conveyed to a TCO substrate and then to the CE.

1.3.3.3 CZTS Solar Cells

Copper Zinc Tin Sulphide (CZTS) solar cells are a contemporary alternative to traditional thin-film solar technologies. The CZTS leverages abundant and non-toxic elements, addressing the toxicity concerns associated with CdTe solar cells. Furthermore, CZTS exhibits near-ideal properties for solar photovoltaics, boasting strong light absorption ($>10^4 cm^{-1}$) and a tunable band gap of 1.0–1.5 eV [23]. Moreover, CZTS(e) material presents benefits including non-toxicity, elemental abundance, and a larger production capacity of its constituents compared to CIGS or CdTe materials [24]. Theoretical calculations suggest efficiencies of approximately 30.9% for CZTS thin-film solar cells. The standard configuration of a conventional CZTS(e) solar cell consists of a soda-lime-glass (SLG) substrate at the bottom, followed by a back-contact layer, absorber layer, buffer layer, window layer, and top contact. Typically, a substrate thickness of approximately 1 mm of SLG is employed for the growth of CZTS(e) solar cells. SLG is selected as the substrate due to its favorable properties for both manufacturing and practical application. The conventional back-contact material utilized for CZTS(e) solar cells is molybdenum (Mo), typically with a thickness of approximately 0.5 mm [25]. Mo is chosen due to its compatible electrical properties and inert behavior towards the CIGS absorber layer. Consequently, Mo serves as the preferred back-contact material for CZTS(e) solar cells, given the similarity in properties between the two absorber layers. Moreover, Mo-coated SLG substrates are readily available in the market.

1.3.3.4 Perovskite Solar Cells

Perovskite materials follow the mineral $CaTiO_3$ structure and it is named after Russian mineralogist L.A. Perovski. Perovskite represents a class of compounds with the ABX_3 crystal structure, where A and B denote cations, and X signifies the anion species [26]. In the context of hybrid organo-lead perovskites, A corresponds to a monovalent organic cation, while B can be Pb(II) or Sn(II), and X typically encompasses halogen anions like I^-, Br^-, or Cl^-. Examples of such hybrid organo-lead perovskites include $CH_3NH_3PbI_3$, $CH_3NH_3PbBr_3$, and mixed halide derivatives. The two primary architectures of a perovskite solar cell are termed mesoscopic and planar and the device structures are shown in Figure 1.7. The components of a PSC structure encompass transparent conducting oxide substrates, an electron transport layer (ETL), a hole transport layer (HTL), and a light-absorbing perovskite layer.

In 2009, Kojima et al. [26] pioneered the development of perovskite material-based solar cells, achieving an efficiency of approximately 3.8%. The Perovskite Solar Cells (PSCs) employ methylammonium lead bromide ($CH_3NH_3PbBr_3$ or $MAPbBr_3$) and methylammonium lead iodide ($CH_3NH_3PbI_3$ or $MAPbI_3$) as absorber layers, featuring bandgaps of 2.3 and 1.6 eV, respectively. Remarkably, in 2021, a PSC with an efficiency of 25.5% marked a significant advancement and showcased competitive performance compared to traditional Silicon-based solar cells [27]. In addition to their application in solar cells, perovskite materials hold potential for use in photodetectors and light-emitting diodes [28,29].

In PSCs, instead of dye materials, perovskite materials like $MAPbI_3$ and $MAPbBr_3$ are utilized as the light absorber. The active layer (perovskite layer) can be deposited using either a one-step (single-step technique) or a two-step (sequential deposition method) approach. In the one-step approach, deposition of the $CH_3NH_3PbI_3$ perovskite film involves the use of lead iodide (PbI_2) and a solvent based on methylammonium iodide (CH_3NH_3I, MAI). Conversely, in the two-step approach, CH_3NH_3I is deposited after PbI_2 to form $CH_3NH_3PbI_3$ perovskite. Typically, when sunlight illuminates a perovskite layer, electrons become excited from the valence band (VB) to the conduction band (CB), leaving holes in the VB. Subsequently, these generated electrons and holes migrate to each interface and are injected into the Electron Transport Material (ETM) and Hole Transport Material (HTM), respectively.

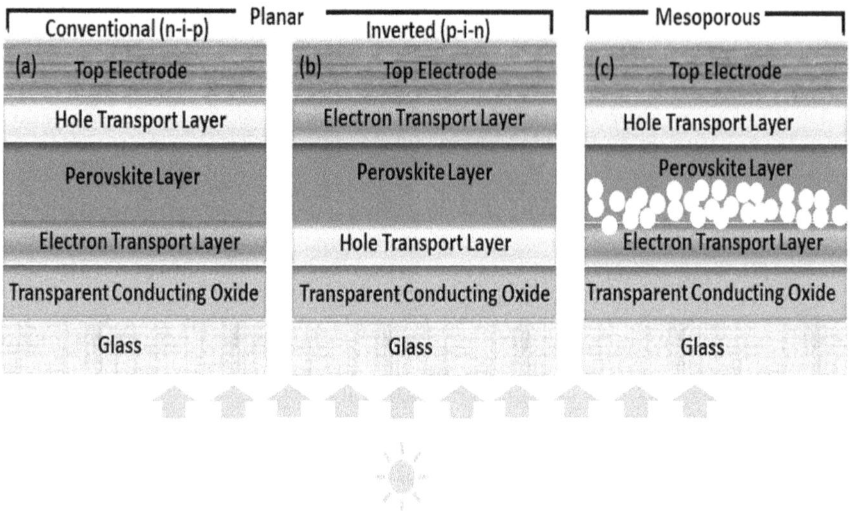

FIGURE 1.7 Perovskite solar cell device structure: (a) conventional, (b) inverted, and mesoporous.

The CB and VB of the perovskite must establish proper energy level alignments with the charge transport layers. Furthermore, charge transport layers characterized by high carrier mobilities are preferred to ensure efficient charge transport. Eventually, electrons and holes are collected by the cathode and anode electrodes, respectively, and are conveyed to the external circuitry for further utilization.

Perovskite materials exhibit remarkable properties that align well with the requirements for solar cell applications. These include a high diffusion length (~1 μm), exceptional carrier mobility, strong light absorption, and low recombination rates. Notably, single-crystalline $MAPbI_3$ demonstrates a defect density of $10^{10} cm^{-3}$, significantly lower than that of polycrystalline perovskites [30].

1.3.3.5 Tin (Sn)-based Perovskite Solar Cells

Among the diverse types of perovskites, Pb^{2+} remains crucial for ensuring exceptional photoelectric properties. However, the byproducts of lead-based perovskites can pose environmental and health risks, hindering the widespread adoption of this technology. Consequently, there has been growing interest in less toxic lead-free perovskite light absorbers. Tin has emerged as a promising alternative to lead due to its lower toxicity and its presence in the same Group 14 (IVA) of the periodic table as lead. Additionally, tin-based perovskites exhibit optical bandgaps ranging from 1.2 to 1.4 eV, closely matching the optimal bandgap (1.34 eV) for the SQ limit under the AM1.5 solar spectrum [31]. Therefore, tin-based perovskites hold significant promise for applications in photovoltaic and other optoelectronic devices. In 2012, a tin-based perovskite was employed as a light absorber for Schottky solar cells, achieving a notable PCE of 0.9% [32]. Since then, tin-based PSCs have experienced rapid advancements, with low-dimensional variants achieving PCEs exceeding 9%. However, a significant challenge for tin-based PSCs lies in their poor air stability due to the facile oxidation of Sn^{2+} to Sn^{4+}. Consequently, various strategies have been devised to prevent the oxidation of tin-based perovskites during device fabrication and operation. Snaith et al. were the first to report complete lead-free Perovskite PSCs based on $MASnI_3$. They adopted a traditional structure with mesoporous TiO_2 as a scaffold, achieving a PCE of 6.40% [33]. Despite notable advancements in this domain, the efficiency of tin-based perovskite solar cells (TPSCs) still falls significantly. Previous investigations have identified several key factors contributing to this performance bottleneck. These include the energy level mismatch at the interface between the ETL and the perovskite, the suboptimal quality of perovskite films arising from their rapid crystallization process, and the susceptibility of Sn^{2+} ions to oxygen [34]. To tackle these challenges, numerous strategies have been devised, primarily involving the incorporation of ammonium salt additives into the precursor solution. These additives interact with the perovskite constituents, regulating the crystallization process and enhancing device performance [35].

1.3.3.6 Organic Solar Cells

Organic Solar Cell (OSC) is a unique type of photovoltaic (PV) cell that utilizes conductive organic polymers or small organic molecules to harness sunlight and generate electricity through the photovoltaic effect. Both small molecules and polymers used in OSCs exhibit large conjugated systems of carbon atoms with alternating single and double bonds. These molecules' electronic structures include a highest occupied molecular orbital (HOMO) and a lowest unoccupied molecular orbital (LUMO), with the separation between them defining the band gap of the material. The device structure is formed by coating the organic semiconductor (donor-acceptor) between two electrodes of different work functions and one of the electrodes should be transparent (Figure 1.8). The donor-acceptor coating can be completed in two ways: firstly coating the donor and acceptor layer-by-layer or mixing the donor and acceptor in a single solution. The coating film using donor-acceptor mixing formed bulk heterojunction (BHJ), and it helped to improve the OSC's photovoltaic parameters. The photogeneration process can be delineated into five distinct steps in OSCs [36]. Initially, a photon is absorbed by one of the bonding electrons in the highest occupied molecular orbital (HOMO), with energy equal to or greater than the donor's or acceptor's bandgap energy (Eg), causing it to transition to the lowest unoccupied molecular orbital (LUMO) and leaving behind a hole. Due to the strong electrostatic attraction and the low dielectric constant of organic semiconductors, bound electron-hole pair states known as excitons are formed. The exciton can diffuse through the material until it reaches the donor/acceptor interface, where it dissociates into free charge carriers. Consequently, the photocurrent generated in the external circuit occurs as free electrons traverse through the acceptor material and holes traverse through the donor. Traditional fullerene derivative acceptors, such as Phenyl-C_{61}-butyric acid methyl ester (PCBM), and Indene-C_{60} Bisadduct (ICBA), played a significant role in the development of OSCs. Due to the fullerenes derivative acceptor's few limitations such as limited band gap tunability, high synthesis cost, morphological instability, etc, the research community switched to polymer and small molecules acceptors, called non-fullerenes acceptors (NFA). The NFA acceptors offer tunability of their band-gaps and energy levels, improve solubility, and enhance donor-acceptor film

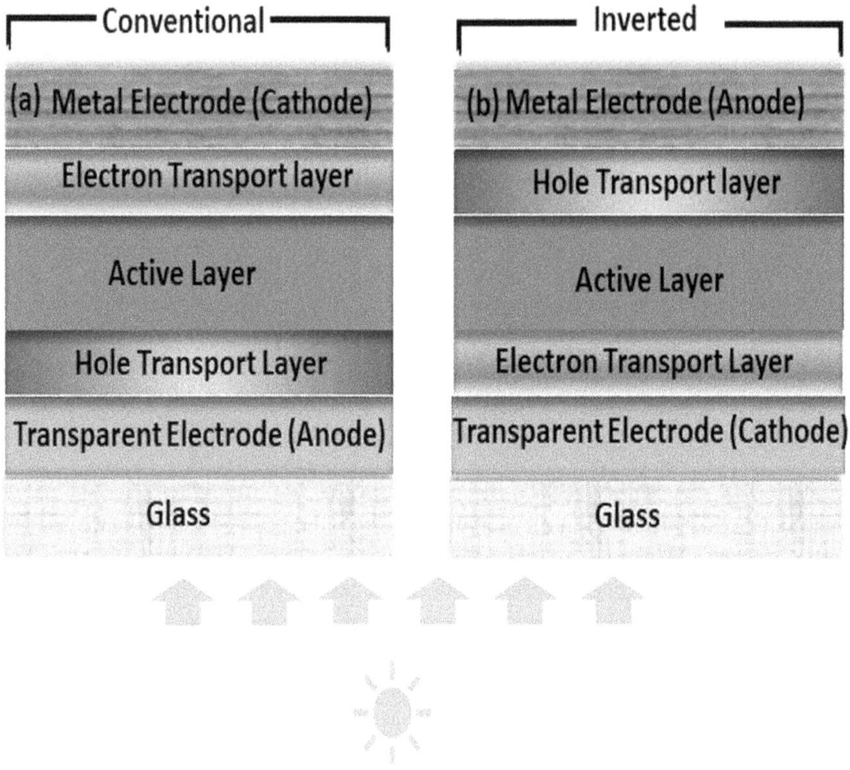

FIGURE 1.8 (a) Conventional and (b) Inverted structures of OSCs.

crystallinity, and with the help of NFA, the OSCs devices PCE reached up to 19% for single junction and 20% for tandem solar cells [37,38].

Figure 1.8 illustrates the two basic geometries of OSCs called conventional and inverted, where the film of organic semiconductor molecules is called the active layer. Over the past few decades, numerous methods have been proposed to enhance the efficiency of OSCs. Improvements in materials and advancements in device structure and engineering have been key areas of focus for enhancing performance. The down-conversion and up-conversion materials have been introduced to utilize the ultra-violet and near-infra-red regions for improving the solar cells' performance, respectively [39,40]. The down-conversion materials absorb the UV region and emit visible light [41,42]. Whereas up-conversion material absorption near-infrared light and emits visible. The concept of ternary and tandem OSCs also have been introduced to improve the PCE by maximizing the absorption of incident solar spectrum. In the ternary, two donors-one acceptor or one donor-two acceptor mixed in a single solution. Whereas, in tandem device structure two or more active layers coated in a single device are joined by intermediate layers. The OSCs unique properties, such as band gap tunability, transparency, and flexibility, making a potential option for broad market applications such as building-integrated photovoltaics, agrivoltaics, indoor photovoltaics, aerospace, underwater, and charging implant devices [43–46].

1.4 APPROACH TO IMPROVE THE POWER CONVERSION EFFICIENCY OF SOLAR CELLS

1.4.1 MULTIJUNTION SOLAR CELLS

Multijunction solar cells use multiple light absorbers (semiconductors) layers in a single device. These solar cells are also known as tandem solar cells. Each light absorber layer has a different band gap that helps to absorb a broader range of incident sunlight. The III-V group materials are used as light absorber layers in multijunction solar cells. In inorganic solar cells, using five junctions, Spectrplab has shown a record-high PCE of 38.8% under AM1.5G (1 Sun) [47]. A similar concept has been introduced in OSCs and perovskite solar cells. In OSCs, different band gap organic semiconductor layers are deposited to broaden the light absorption. Recently, OSCs delivered 20.6% PCE using the tandem device structure [38]. Perovskite solar cells play a significant role in making multiple solar cells. Perovskite-perovskite, perovskite-organic, and perovskite-inorganic multijunction solar cells show a significant improvement in PCE compared to their single-junction solar cells. Perovskite-organic tandem solar cells improved the PCE after adding perovskite solar cells by 14.43% (single junction) and organic solar cells by 14.01% (single junction), and the tandem solar cells delivered PCE of 18.06% [48]. Perovskite materials are compatible with inorganic solar cells such as Silicon, CIGS, GaAs, and CdTe for making tandem solar cells.

1.4.2 CONCENTRATING SOLAR CELLS

Concentrating Solar cells (CSCs) utilize optical devices to focus a large area of sunlight onto a solar cell. The preliminary research on CSC technology started in the 1970s and this technology has become a viable alternative in the solar electricity sector. There is no dominant design due to its relative youth. CSC systems are best suited for areas with high direct normal irradiance, as they can only utilize direct beam radiation and not diffuse radiation [49]. CSC systems use optics to focus solar energy onto high-efficiency multi-junction solar cells. Originally designed for space applications, these cells have been adapted for terrestrial concentrator solar applications. By pairing small cells with less expensive optics, CSC systems retain superior performance while reducing the amount of expensive semiconductor material. CSC systems can significantly increase surface temperatures on solar cell materials, necessitating equal distribution of energy to avoid local overheating. Active or passive cooling procedures may be used to manage temperature, with high-concentration cells requiring high-capacity heat sinks to prevent thermal destruction. Concentration ratios for <10, 10–100, 100–2000, and >2000 suns are called low, medium, high, and ultra-high, respectively [50]. The Fresnel lens is considered the most promising for concentrated solar cell applications. This technology offers several advantages. First of all, It requires less photovoltaic material to capture the same sunlight compared to non-concentrating solar cell systems. It enables high power generation and improves the use of expensive multi-junction cells economically, due to smaller space requirements. However, concentrating light requires direct sunlight rather than diffuse light, limiting its effectiveness to clear, sunny locations and necessitating tracking in most cases.

1.5 CONCLUSION

The concluding remarks of the chapter on solar cell generations underscore the remarkable progress achieved in solar cell technology, spanning from the inception of the first generation to the promising emergence of the third generation. This progression highlights the evolution from conventional crystalline silicon cells to groundbreaking innovations like thin film, polymer-based, and perovskite-based cells. Each generational leap brings forth distinct characteristics, benefits, and applications, significantly influencing the trajectory of solar energy utilization. The initial phase of solar cell development, known as the first generation, lays the groundwork with its reliance on crystalline silicon. This category encompasses various types, such as mono-crystalline, and poly-crystalline silicon cells, which have long dominated the industry owing to their reliability and efficiency. Transitioning to the second generation marks a shift towards thin film technology, offering increased flexibility and cost-effectiveness compared to traditional crystalline silicon cells. Notable examples within this generation include amorphous silicon, Copper Indium Gallium Di-Selenide (CIGS), and Cadmium Telluride cells, which expand the horizons of solar integration across diverse environments and applications.

The advent of the third generation introduces a diverse array of materials and designs to enhance efficiency further, reduce costs, and broaden the scope of solar energy utilization. This generation encompasses polymer-based cells, dye-sensitized cells, nano-crystal-based cells, CZTS cells, Quantum Dot cells, and Perovskite cells. Organic and perovskite solar cells stand out for their exceptional efficiency and potential for large-scale, cost-effective production. Multi-junction and concentrating solar cells represent a promising approach to improving the solar cells' PCE. Concentrating solar cells utilizing optical devices to focus sunlight onto small-size solar cell devices. It offers advantages such as reduced material usage and cost-effective, high-efficiency cell utilization. However, their effectiveness is limited to areas with high direct normal irradiance, and careful heat management is essential to maintain optimal performance and long-term stability. Furthermore, the discourse explores specific advancements within the third generation, such as tin-based and carbon-based perovskite solar cells, underscoring ongoing research and development endeavors to surmount existing challenges and unlock the full potential of these cutting-edge technologies.

In summary, the conclusion emphasizes the perpetual evolution of solar cell technology. It underscores the pivotal role of continued research and development in propelling further enhancements and adoption of renewable energy sources. Through the embrace of diverse materials and innovative designs, solar cells hold the promise of revolutionizing energy production and steering humanity toward a sustainable future.

NOTE

* Corresponding author: Vinay Gupta, Department of Physics, Khalifa University, Abu Dhabi, United Arab Emirates.

REFERENCES

1. Y. Huang, Z. Kuldasheva, S. Bobojanov, B. Djalilov, R. Salahodjaev, and S. Abbas, "Exploring the links between fossil fuel energy consumption, industrial value-added, and carbon emissions in G20 countries," *Environmental Science and Pollution Research*, vol. 30, no. 4, pp. 10854–10866, Sep. 2022, doi: 10.1007/s11356-022-22605-9
2. T.-Z. Ang, M. Salem, M. Kamarol, H. S. Das, M. A. Nazari, and N. Prabaharan, "A comprehensive study of renewable energy sources: Classifications, challenges and suggestions," *Energy Strategy Reviews*, vol. 43, p. 100939, Sep. 2022, doi: 10.1016/j.esr.2022.100939
3. M. A. Green, E. D. Dunlop, J. Hohl-Ebinger, M. Yoshita, N. Kopidakis, and X. Hao, "Solar cell efficiency tables (Version 58)," *Progress in Photovoltaics: Research and Applications*, vol. 29, no. 7, pp. 657–667, Jul. 2021, doi: 10.1002/pip.3444
4. W. Shockley and H. J. Queisser, "Detailed balance limit of efficiency of *p-n* junction solar cells," *Journal of Applied Physics*, vol. 32, no. 3, pp. 510–519, Mar. 1961, doi: 10.1063/1.1736034
5. A. H. Alami, S. Alasad, H. Aljaghoub, M. Ayoub, A. Alashkar, A. Mdallal, R. Hasan, "Second-generation photovoltaics: Thin-film technologies," In *Advances in Science, Technology and Innovation* (pp. 65–75). Springer Nature, 2023. https://link.springer.com/chapter/10.1007/978-3-031-31349-3_6
6. NREL's, "Best-research-cell-efficiencies," Accessed: Apr. 23, 2024. [Online]. Available: https://www.nrel.gov/pv/interactive-cell-efficiency.html
7. T. Niewelt et al., "Taking monocrystalline silicon to the ultimate lifetime limit," *Solar Energy Materials and Solar Cells*, vol. 185, pp. 252–259, Oct. 2018, doi: 10.1016/j.solmat.2018.05.040
8. T. Saga, "Advances in crystalline silicon solar cell technology for industrial mass production," *Nature Publishing Group Asia Materials*, vol. 2, no. 3, pp. 96–102, Jul. 2010, doi: 10.1038/asiamat.2010.82
9. R. C. Chittick, J. H. Alexander, and H. F. Sterling, "The preparation and properties of amorphous silicon," *Journal of the Electrochemical Society*, vol. 116, no. 1, p. 77, 1969, doi: 10.1149/1.2411779
10. A. V. Shah et al., "Thin-film silicon solar cell technology," *Progress in Photovoltaics: Research and Applications*, vol. 12, no. 2–3, pp. 113–142, 2004, doi: 10.1002/pip.533
11. J. Ramanujam and U. P. Singh, "Copper indium gallium selenide based solar cells – A review," *Energy & Environmental Science*, vol. 10, no. 6, pp. 1306–1319, Jun. 2017, doi: 10.1039/C7EE00826K
12. P. Jackson, R. Wuerz, D. Hariskos, E. Lotter, W. Witte, and M. Powalla, "Effects of heavy alkali elements in $Cu(In,Ga)Se_2$ solar cells with efficiencies up to 22.6%," *Physica Status Solidi (RRL) – Rapid Research Letters*, vol. 10, no. 8, pp. 583–586, Aug. 2016, doi: 10.1002/pssr.201600199
13. M. A. Steiner, J. F. Geisz, I. García, D. J. Friedman, A. Duda, and S. R. Kurtz, "Optical enhancement of the open-circuit voltage in high-quality GaAs solar cells," *Journal of Applied Physics*, vol. 113, no. 12, Mar. 2013, doi: 10.1063/1.4798267
14. J. F. Geisz, M. A. Steiner, I. García, S. R. Kurtz, and D. J. Friedman, "Enhanced external radiative efficiency for 20.8% efficient single-junction GaInP solar cells," *Applied Physics Letters*, vol. 103, no. 4, Jul. 2013, doi: 10.1063/1.4816837
15. K. Makita et al., "Mechanical stacked GaAs//$CuIn1_{-y}Ga_ySe_2$ three-junction solar cells with 30% efficiency via an improved bonding interface and area current-matching technique," *Progress in Photovoltaics: Research and Applications*, vol. 31, no. 1, pp. 71–84, Jan. 2023, doi: 10.1002/pip.3609
16. N. S. Noorasid et al., "Current advancement of flexible dye-sensitized solar cell: A review," *Optik (Stuttg)*, vol. 254, p. 168089, Mar. 2022, doi: 10.1016/j.ijleo.2021.168089

17. K. Kakiage, Y. Aoyama, T. Yano, K. Oya, J. Fujisawa, and M. Hanaya, "Highly-efficient dye-sensitized solar cells with collaborative sensitization by silyl-anchor and carboxy-anchor dyes," *Chemical Communications*, vol. 51, no. 88, pp. 15894–15897, Sep. 2015, doi: 10.1039/C5CC06759F
18. N. Sekar and V. Y. Gehlot, "Metal complex dyes for dye-sensitized solar cells: Recent developments," *Resonance*, vol. 15, no. 9, pp. 819–831, Sep. 2010, doi: 10.1007/s12045-010-0091-8
19. K. Prajapat, M. Dhonde, K. Sahu, P. Bhojane, V. Murty, and P. M. Shirage, "The evolution of organic materials for efficient dye-sensitized solar cells," *Journal of Photochemistry and Photobiology C: Photochemistry Reviews*, vol. 55, p. 100586, Jun. 2023, doi: 10.1016/j.jphotochemrev.2023.100586
20. S. S. Bopardekar, K. Chandrasekhar, and N. R. Patil, "Synthesis, structural and optical investigation of Cdse semiconductor quantum dots," *Materials Today Proceedings*, vol. 24, pp. A1–A5, 2020, doi: 10.1016/j.matpr.2020.04.451
21. J. M. Luther, J. Gao, M. T. Lloyd, O. E. Semonin, M. C. Beard, and A. J. Nozik, "Stability assessment on a 3% bilayer PbS/ZnO quantum dot heterojunction solar cell," *Advanced Materials*, vol. 22, no. 33, pp. 3704–3707, Sep. 2010, doi: 10.1002/adma.201001148
22. M. Hao et al., "Ligand-assisted cation-exchange engineering for high-efficiency colloidal $Cs_{1-x}FA_xPbI_3$ quantum dot solar cells with reduced phase segregation," *Nature Energy*, vol. 5, no. 1, pp. 79–88, Jan. 2020, doi: 10.1038/s41560-019-0535-7
23. M. Kangsabanik and R. N. Gayen, "A comprehensive review on the recent strategy of cation substitution in CZTS(Se) thin films to achieve highly efficient kesterite solar cells," *Solar RRL*, vol. 7, no. 24, Dec. 2023, doi: 10.1002/solr.202300670
24. H. S. Nugroho et al., "A progress review on the modification of CZTS(e)-based thin-film solar cells," *Journal of Industrial and Engineering Chemistry*, vol. 105, pp. 83–110, Jan. 2022, doi: 10.1016/j.jiec.2021.09.010
25. S. Zhang et al., "Band alignment tuning at Mo/CZTS back contact interface through surface oxidation states control of Mo substrate," *Solar Energy Materials and Solar Cells*, vol. 229, p. 111141, Aug. 2021, doi: 10.1016/j.solmat.2021.111141
26. A. Kojima, K. Teshima, Y. Shirai, and T. Miyasaka, "Organometal halide perovskites as visible-light sensitizers for photovoltaic cells," *Journal of the American Chemical Society*, vol. 131, no. 17, pp. 6050–6051, May 2009, doi: 10.1021/ja809598r
27. P. Basumatary and P. Agarwal, "A short review on progress in perovskite solar cells," *Materials Research Bulletin*, vol. 149, p. 111700, May 2022, doi: 10.1016/j.materresbull.2021.111700
28. K. Zhang, N. Zhu, M. Zhang, L. Wang, and J. Xing, "Opportunities and challenges in perovskite LED commercialization," *Journal of Materials Chemistry C*, vol. 9, no. 11, pp. 3795–3799, Mar. 2021, doi: 10.1039/D1TC00232E
29. X. Huang, Y. Guo, and Y. Liu, "Perovskite photodetectors and their application in artificial photonic synapses," *Chemical Communications*, vol. 57, no. 87, pp. 11429–11442, Nov. 2021, doi: 10.1039/D1CC04447H
30. F. Wang, S. Bai, W. Tress, A. Hagfeldt, and F. Gao, "Defects engineering for high-performance perovskite solar cells," *Nature Partner Journal: Flexible Electronics*, vol. 2, no. 1, p. 22, Aug. 2018, doi: 10.1038/s41528-018-0035-z
31. M. Pitaro, E. K. Tekelenburg, S. Shao, and M. A. Loi, "Tin halide perovskites: From fundamental properties to solar cells," *Advanced Materials*, vol. 34, no. 1, Jan. 2022, doi: 10.1002/adma.202105844
32. Z. Chen, J. J. Wang, Y. Ren, C. Yu, and K. Shum, "Schottky solar cells based on $CsSnI_3$ thin-films," *Applied Physics Letters*, vol. 101, no. 9, p. 093901, Aug. 2012, doi: 10.1063/1.4748888
33. N. K. Noel et al., "Lead-free organic-inorganic tin halide perovskites for photovoltaic applications," *Energy & Environmental Science*, vol. 7, no. 9, pp. 3061–3068, 2014, doi: 10.1039/C4EE01076K

34. J. Wang et al., "Controlling the crystallization kinetics of lead-free tin halide perovskites for high performance green photovoltaics," *Advanced Energy Materials*, vol. 11, no. 39. John Wiley and Sons Inc, Oct. 01, 2021. doi: 10.1002/aenm.202102131
35. X. Meng et al., "Highly stable and efficient $FASnI_3$-based perovskite solar cells by introducing hydrogen bonding," *Advanced Materials*, vol. 31, no. 42, Oct. 2019, doi: 10.1002/adma.201903721
36. S. Bishnoi et al., "Engineered cathode buffer layers for highly efficient organic solar cells: A review," *Advanced Materials Interfaces*, vol. 9, no. 19. John Wiley and Sons Inc, Jul. 01, 2022. doi: 10.1002/admi.202101693
37. L. Zhu et al., "Single-junction organic solar cells with over 19% efficiency enabled by a refined double-fibril network morphology," *Nature Materials*, vol. 21, no. 6, pp. 656–663, Jun. 2022, doi: 10.1038/s41563-022-01244-y
38. Z. Zheng et al., "Tandem organic solar cell with 20.2% efficiency," *Joule*, vol. 6, no. 1, pp. 171–184, Jan. 2022, doi: 10.1016/j.joule.2021.12.017
39. R. Datt et al., "Down-conversion materials for organic solar cells: Progress, challenges, and perspectives: Photovoltaics: Special issue dedicated to professor Yongfang Li," *Aggregate*, vol. 3, no. 3. John Wiley & Sons Inc, Jun. 01, 2022. doi: 10.1002/agt2.185
40. B. Ehrler, N. Yanai, and L. Nienhaus, "Up- and down-conversion in molecules and materials," *Journal of Chemical Physics*, vol. 154, no. 7. AIP Publishing, LLC, 2021. doi: 10.1063/5.0045323
41. S. Bishnoi et al., "Luminescent $GdVO_4{:}Sm^{3+}$ quantum dots enhance the power conversion efficiency of bulk heterojunction polymer solar cells by Förster resonance energy transfer," *Applied Physics Letters*, vol. 109, no. 2, Jul. 2016, doi: 10.1063/1.4958690
42. S. Bishnoi, V. Gupta, C. Sharma, D. Haranath, M. Kumar, and S. Chand, "Luminescent cathode buffer layer for enhanced power conversion efficiency and stability of bulk-heterojunction solar cells," *Organic Electronics*, vol. 38, pp. 193–199, Nov. 2016, doi: 10.1016/j.orgel.2016.08.017
43. Y. Yang, C. Xue, H. Yin, Z. Chen, and X. T. Hao, "Bandgap matching strategy for organic photovoltaic cells in oceanic applications," *Cell Reports Physical Science*, vol. 3, no. 5, May 2022, doi: 10.1016/j.xcrp.2022.100861
44. R. Datt, H. K. H. Lee, G. Zhang, H. L. Yip, and W. C. Tsoi, "Organic solar cells at stratospheric condition for high altitude platform station application†," *Chinese Journal of Chemistry*, vol. 40, no. 24, pp. 2927–2932, Dec. 2022, doi: 10.1002/cjoc.202200481
45. R. Datt, H. K. H. Lee, M. Spence, M. Carnie, and W. C. Tsoi, "High performance non-fullerene organic photovoltaics under implant light illumination region," *Applied Physics Letters*, vol. 122, no. 14, Apr. 2023, doi: 10.1063/5.0144861
46. X. Hu et al., "Facile synthesis of annulated benzothiadiazole derivatives and their application as medium band gap acceptors in organic photovoltaic devices," *Journal of Materials Chemistry C*, vol. 10, no. 24, pp. 9249–9256, Jun. 2022, doi: 10.1039/d2tc01433e
47. H. Sugimoto, High efficiency and large volume production of CIS-based modules. *40th IEEE Photovoltaic Specialists Conference (PVSC)*, 2014: Colorado Convention Center, Denver, Colorado, 8–13 June 2014.
48. X. Wu et al., "Improved stability and efficiency of perovskite/organic tandem solar cells with an all-inorganic perovskite layer," *Journal of Materials Chemistry A*, vol. 9, no. 35, pp. 19778–19787, Sep. 2021, doi: 10.1039/d0ta12286f
49. T. S. Viana, R. Rüther, F. R. Martins, and E. B. Pereira, "Assessing the potential of concentrating solar photovoltaic generation in Brazil with satellite-derived direct normal irradiation," *Solar Energy*, vol. 85, no. 3, pp. 486–495, Mar. 2011, doi: 10.1016/j.solener.2010.12.015
50. K. Shanks, S. Senthilarasu, and T. K. Mallick, "Optics for concentrating photovoltaics: Trends, limits and opportunities for materials and design," *Renewable and Sustainable Energy Reviews*, vol. 60. Elsevier Ltd, pp. 394–407, Jul. 01, 2016. doi: 10.1016/j.rser.2016.01.089

2 Components, Working, Parameters, and Characteristics of Solar Cells

*Shivani Dhall, Shilpa Setia, Summit Jalota, Jyoti Prakash, and Kapil Sood**

2.1 INTRODUCTION TO SOLAR CELL

The reservoir of fossil fuels is incapable of meeting our requirements in the coming decades. Our daily needs are entirely dependent on energy, with the primary sources being petroleum and gas, both of which are decreasing exponentially and last for only a couple of decades. The increasing dependence will also lead to rising prices, and in the future, prices will likely increase much faster than they do today. The increasing population has exponential energy demand which is met from fossil fuels which further results in an increase in the emission of CO_2 and other pollutants. All these pollutants are endangered by environmental changes. Therefore, there is an urgent need to use alternative/renewable and inexpensive sources of energy such as wind, solar, bioenergy (organic matter burned as a fuel), and hydroelectric, tides. All these forms of energy are available in plenty and will not run out. Particularly, in the present world, the use of solar is considered to be the best and most inexpensive alternative to fossil fuels. Solar energy is an unlimited source of energy that is available free of cost. Due to this reason, solar cells have gained much attention in the scenario of energy in comparison to other alternative sources.

The primary mechanism for converting solar energy into electrical energy in solar cells is called the photovoltaic effect. Firstly, Bell Labs made the silicon (Si) based solar cell in 1954 using the photovoltaic effect [1], afterwards, constant efforts have been put by the scientific community to find the most efficient and cost-effective solar cells so that the world does not remain dependent on fossil fuels. In this direction, many kinds of solar cells have been made using a variety of materials. In order to be efficient, solar cells need to absorb a particular range of energy from the solar spectrum. The range of the solar spectrum lies between 100 nm and 1 mm. But mostly, the irradiance occurs between 250 and 2500 nm with maxima in the visible range of 400–700 nm. This means that solar cells should absorb in the visible range to get maximum efficiency [2].

A group of solar cells, which are the smallest units of a photovoltaic system, are soldered together to produce a solar panel (Figure 2.1).

 DOI: 10.1201/9781003499701-2

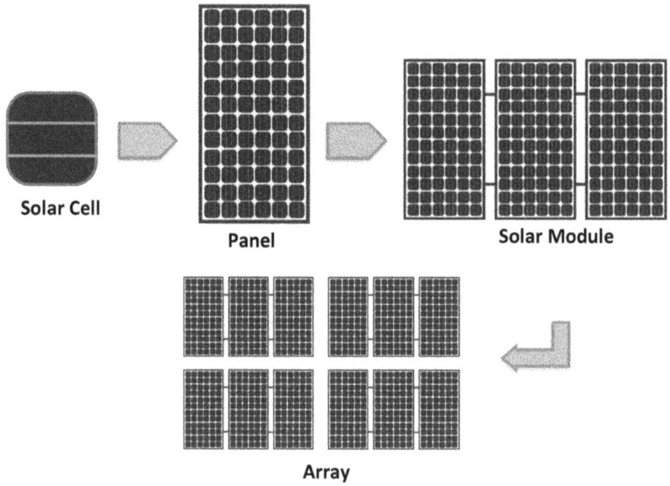

FIGURE 2.1 Hierarchy of solar cell to the solar array.

A plastic sheet is attached at the back and a glass sheet is attached at the front of these solar panels. After that, a solar module is formed by using a lot of these solar panels. A photovoltaic array is the complete power-generating unit, consisting of several PV modules and panels. Currently, most of the solar panels that are available in the market use Si as the base of photovoltaic material. These solar panels are large and are very heavy. So, the transportation of these solar panels is difficult. Also, Si-based solar cells are expensive and have low efficiency. In this direction, direction, numerous research is going on to generate solar cells with low-cost materials and high efficiency. In this chapter, we have discussed the components and workings of these different types of solar cells along with their parameters.

2.2 MAIN COMPONENTS OF SOLAR PANEL

The primary components of a solar panel (Figure 2.2) include: (a) top glass, (b) anti-reflective coating, (c) solar cell, (d) encapsulant, (e) back sheet, (f) junction box and (g) frame [3].

a. Top glass: It is the outermost layer of the solar cell. It is generally made of tempered glass of 3–4 mm thickness. Tempered glass is used because it can withstand impact and heat stress. This glass protects photovoltaic cells from external factors such as rain, dust, salts, and gases. This glass should be transparent so as to allow maximum sunlight to enter the cell.
b. Anti-reflective coating: On the top glass, anti-reflective coating is applied. This coating helps to increase the amount of sunlight absorbed by the solar cell. This will be discussed in detail later in this chapter.
c. Solar cell: Under the glass, there are little grid patterns with slight metallic strips that act as the top electrode of the solar cell. Photovoltaic material

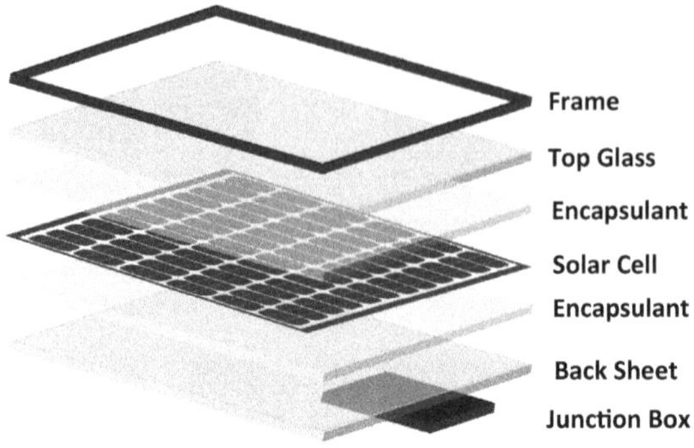

FIGURE 2.2 The components of a solar cell.

is the heart of solar cells since it allows electricity to be produced via the photovoltaic effect. This is present beneath this top electrode. Solar cells use a variety of photovoltaic materials as explained in upcoming sections. The backside of the solar cell contains a continuous layer of electrodes that work with the metallic grid (top electrode) to generate an electric current. Different types of solar cells use different materials as back electrodes, as discussed in upcoming sections.

d. Encapsulant: Encapsulant material is used in solar cells to bind various layers, as shown in Figure 2.1. Encapsulants are transparent, weather and temperature-resistant materials that increase the durability and life of solar cells. Most commonly used encapsulant is polymer ethylene vinyl acetate (EVA) due to its low cost. This is available in the market in the form of rolls. It is cut into sheets before applying. It is translucent but becomes transparent after heating. The main drawback of EVA is it gets hydrolyzed in the presence of light or oxygen to form acetic acid. Acetic acid can corrode solar cells. Also, it becomes yellow after some time. Another material that can be used in place of EVA is polyolefin (POE). It has better properties as compared to EVA. As it is costlier than EVA, it is not used commonly.
e. Back sheets: They also act as a protective layer. They are available as white, black, or transparent sheets. These sheets are made up of polymers such as polyvinyl fluoride, ethylene tetrafluoroethylene, polyvinylidene fluoride and polyester.
f. Junction box: It is installed at the back of the solar panel. Its main purpose is to protect the solar panel wiring from harsh conditions.
g. Frame: The frame has multiple objectives, such as protection against adverse weather conditions, protection against mechanical damage and providing the optimal angle for mounting the solar panel. For rooftop installations, an aluminum frame is used, and for ground installations, steel or aluminum can be used. Sometimes, a sealing tape or a silicon sealant is also used along with a frame to protect the cell from moisture.

2.2.1 Role of Antireflective Coating

An antireflective coating (ARC) is specially designed to reduce the light reflected away from the solar cell. The efficiency of solar cells decreases due to the loss of light owing to an optical phenomenon known as reflection from the surface [4]. This fraction of reflected light (neglecting the scattering and absorption effect as the contribution is very small) is subtracted from the light transmitted to the cell. As the number of reflections increases, efficiency decreases. To overcome this issue, dielectric and metallic thin films are coated over the surface of the solar cell, known as an antireflective coating. This increment in transmission light elevates the absorption by the semiconductor material, thereby increasing the efficiency in the generation of electricity. ARCs are synthesized to minimize the Fresnel reflection losses that occur when light propagates through interfaces between materials with different refractive indices [5]. The objective underlying ARCs is to induce destructive interference between reflected waves at the interface, therefore reducing the amount of light reflected and increasing the transmitted light.

To understand this concept, consider a single dielectric thin film (antireflective) with refractive index, n, which grows on a semiconductor having refractive index n_s such that $n < n_s$. This thin film has two interfaces (air-thin film and thin film-semiconductor). A thin film acts as an ARC if the waves reflected from the upper and lower surfaces of the coated film are in opposite phases so that these reflected rays from these two interfaces experience destructive interference, which causes the reflected rays to cancel each other out, resulting in improved transmission. The first condition for this destructive interference is the phase change of π (opposite phase) occurs when light is reflected from a medium with a higher refractive index compared to the medium through which the light is traveling. Another condition is the thickness t of the antireflective film, which should be one-fourth of the wavelength of the incident ray. Figure 2.3 demonstrates the destructive interference in the case of a layer of thin film on a semiconductor.

In order to compute the reflectance r at normal incidence, the following formula is employed:

$$r = \left[\frac{n_{\text{air}} n_s - n^2}{n_{\text{air}} n_s + n^2}\right]^2 \tag{2.1}$$

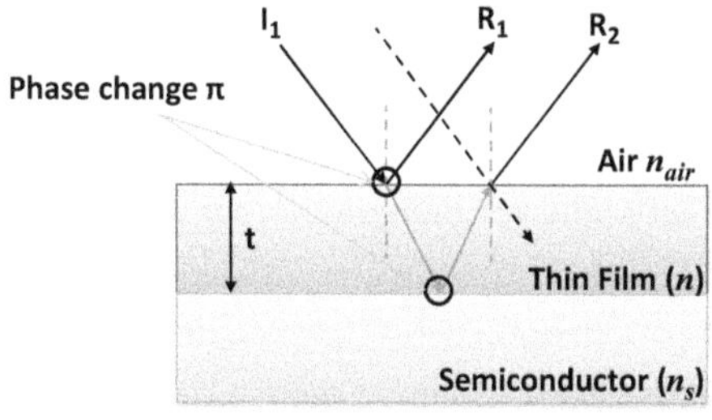

FIGURE 2.3 Illustration of the destructive interference in case of single layer antireflective thin film on semiconductor.

Where n_{air} is the reflectance in air, n_s is reflectance for semiconductor materials and n is reflectance for antireflective film [6].

Single-layer coatings are appropriate for limited wavelength bands, whereas multilayer thin film coatings are better at covering a wide range of wavelengths. Furthermore, multilayer coatings perform better at different angles of incidence and maintain their optical characteristics with relatively little variation. Detailed analysis of the optical and mechanical properties of various materials is crucial for obtaining effective ARCs with outstanding performances.

The conditions for a good anti-reflection coating in a solar cell should include a low index of refraction and high transmissivity, combined with a thin film chemistry that is resistant to degradation and humidity. The coating should be applied uniformly and in a way that minimizes surface roughness and defects, which can also contribute to anti-reflective losses. In addition, the regular cleaning and maintenance of the solar cell surface can also help to ensure that the anti-reflective coating remains effective. The optical and physical properties of the coating must remain stable under long-term operation at elevated temperatures, repeated thermal cycling, UV radiations, and other such conditions [7]. Moreover, single-layer anti-reflective coating (SLARC) and double-layer anti-reflective coating reduces reflectivity only in a limited range. However, triple layer anti-reflective coatings (TLARC) significantly improve the conversion efficiency and current density in Si-based solar cells. For optimal optical alignment of the solar cell, concerning incident sunlight, such a triple-layer anti-reflective coating can eliminate the need for a mechanical tracking mechanism.

2.3 WORKING OF SOLAR CELLS

It is well known that based on the photovoltaic material, solar cells can be divided into four categories: First, second, third, and fourth generation. First-generation solar cells are mainly made up of Si. Second-generation solar cells are made up of cadmium telluride (CdTe), copper indium gallium selenide (CIGS), amorphous silicon, and copper zinc tin sulfide (CZTS). In third-generation solar cells, a variety of new materials such as inorganic nanoparticles, organic molecules, perovskite, dyes, carbon nanotubes, quantum dots, and many more are being used [8]. Fourth-generation solar cells contain molybdenum disulphide (MoS_2), graphene, tungsten disulphide (WS_2), or tungsten diselenide (WSe_2) etc., in the central layer. In order to provide more insight into the working principles of solar cells, Liu et al. [9] categorized these cells further into three categories based on junctions applied: (a) inorganic p-n junction–based solar cells, (b) organic e-donor/e-acceptor junction–based solar cells and (c) inorganic-organic hybrid junction–based solar cells as shown in Figure 2.4.

2.3.1 p-n Junction-Based Cells

In general, the working of an inorganic p-n junction based solar cell can be summarized as:

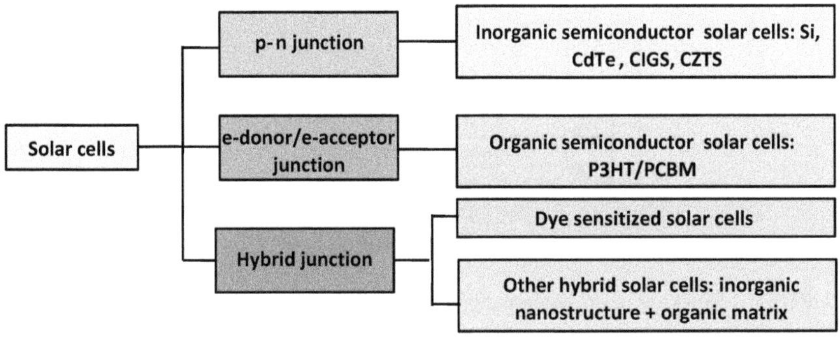

FIGURE 2.4 Classification of solar cells based on junctions applied. CIGS (copper indium gallium selenide); CZTS (copper zinc tin sulfide); P3HT (poly(3-hexylthiophene)); PCBM ([6,6]-phenyl C_{61}-butyric acid methylester).

a. When photons reach the p-n junction, they penetrate into a thin p-doped layer and transfer their energy to the semiconductor material. Electron-hole pairs are created in the depletion region and finally, produced free electrons that move to the n-doped side of the junction.
b. Holes in the depletion zone swept towards the p-doped side of the junction.
c. When newly created holes are driven to the p-doped side, they encounter the barrier potential at the junction, which prevents them from further crossing over. The concentration of electrons becomes higher on the n-doped side of the junction, while the concentration of holes becomes greater on the p-doped side. This difference in charge carrier concentrations establishes a voltage, known as the photovoltage, effectively causing the p-n junction to behave like a small battery cell. Furthermore, if a small load is connected across the junction, a tiny current will flow through it due to the establishment of the photovoltage [10].

In Si-based solar cells, the top electrode is usually a metallic grid made of either silver or aluminum. The back electrode is also made up of either silver or aluminum. This solar cell contains two layers: p-doped and n-doped. The p-doped layer is made by doping Si with boron or gallium, which have one less electron than is required to form bonds with surrounding Si atoms. This results in the formation of holes. Similarly, the n-doped layer is made by doping Si with phosphorus, which has one extra electron, required to form bonds with surrounding Si atoms. This results in the formation of free electrons [11]. The basic working for the first-generation Si solar cells is shown in Figure 2.5. The Shockley-Queisser limit indicates that maximum solar conversion efficiency occurs in a material with a band gap of 1.34 eV [12]. So, the materials with band gaps lower than 1.34 eV can easily generate electricity from lower-energy photons. As the band gap of Si is 1.1 eV, it enables broad absorption of solar radiation. Since this bandgap is lower than 1.34 eV, energy losses take place when high-energy photons are absorbed.

Cadmium telluride (CdTe) is a direct-band gap material with a bandgap value of 1.4–1.5 eV and has p-type conductivity [13,14]. A p-n heterojunction in CdTe-based solar cells is made up of an n-type CdS layer deposited on top of a p-type CdTe layer. Cadmium sulfide is covered by a transparent conductive front contact, which is often

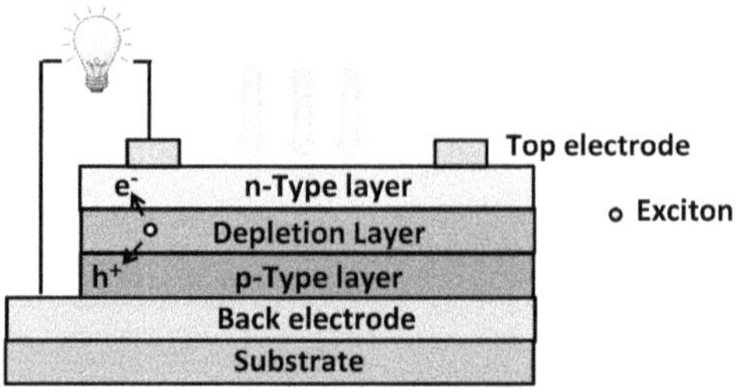

FIGURE 2.5 Working of an inorganic p-n junction-based solar cell in general.

fluorine-doped tin oxide (SnO_2:F) [8,15]. The metal layer is placed on the back (in contact with the CdTe layer) to form electrical contacts, as shown in Figure 2.5. In copper indium gallium selenide (CIGS) cells, the bandgap varies from approximately 1.0eV (for copper indium selenide) to about 1.7eV (for copper gallium selenide). So, by changing the amount of indium/gallium, the band gap can be tuned to optimize the performance. p-n heterojunction of n-type CdS on top of a p-type layer of CIGS is made [16]. CZTS (copper zinc tin sulfide) has a band gap of ~1.4eV and exhibits p-type conductivity. In these solar cells, heterojunction is made of p-type CZTS and n-type CdS layers [17]. In both CIGS and CZTS cells, Al-doped ZnO serves as a transparent top electrode, while Mo is used to form a back electrode.

A new class of solar cells known as HIT (Heterojunction with Intrinsic Thin Layer) utilizes two kinds of silicon as p-type and n-type layers. This solar cell is composed of a p–i–n–i–n-doped stack of silicon layers as shown in Figure 2.6. The central layer contains an n-type crystalline silicon wafer and the others are made of amorphous thin layers of Si. The electrodes are made of a transparent conducting oxide (TCO) and metal grid [18].

2.3.2 e-Donor e-Acceptor Junction-Based Cells

Organic solar cells are composed of small organic or polymeric materials. Organic solar cells are less expensive than other forms of solar cells. Organic or polymer solar cells were engineered to make a more versatile and flexible solar cell. In the case of organic solar cells, a band gap refers to the difference in energies between the

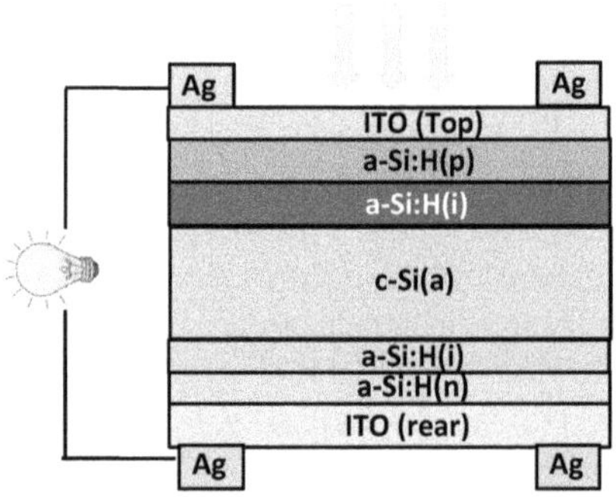

FIGURE 2.6 Working of heterojunction with intrinsic thin layer solar cells.

HOMO (Highest Occupied Molecular Orbital) and the LUMO (Lowest Unoccupied Molecular Orbital). Typically, organic solar cells have a band gap that ranges from 2.0 to 2.5 eV, which allows them to absorb visible and near-infrared light. Both conjugated polymer and small organic molecules can be utilized in organic solar cells, and the band gap in these materials can be tailored by changing factors such as bond length, resonance energy, planarity, substituent, and conjugation [15].

These cells can either have a single layer structure, a bilayer structure, or a bulk-heterojunction structure, as shown in Figure 2.7. In the case of single-layer structure, the organic layer is sandwiched between two metallic conductors, typically a layer of indium tin oxide (ITO) with high work function and a layer of low work function metal such as aluminum, magnesium, or calcium. When sunlight falls on the device, the organic molecule absorbs light, and excitons (electron-hole pairs) are created. The different work functions of two conductors create a potential that helps to split the exciton pairs, transporting electrons toward the positive electrode and holes toward the negative electrode. But these devices have very low efficiency. Such solar cells incorporate materials like magnesium phthalocyanine and polyacetylene [19–21].

In the bilayer structure, two organic materials are utilized. To create a donor–acceptor heterojunction, one of them takes the role of an electron donor and the other as an acceptor. Upon absorption of photons from sunlight, an electron-hole pair (exciton) is generated in the donor material, which then diffuses to the donor-acceptor interface. Excitons split into free electrons and holes by the effective field formed when two different materials form a heterojunction. Also, electrons fall from the LUMO of the donor to the LUMO of the acceptor. At the same time, when an

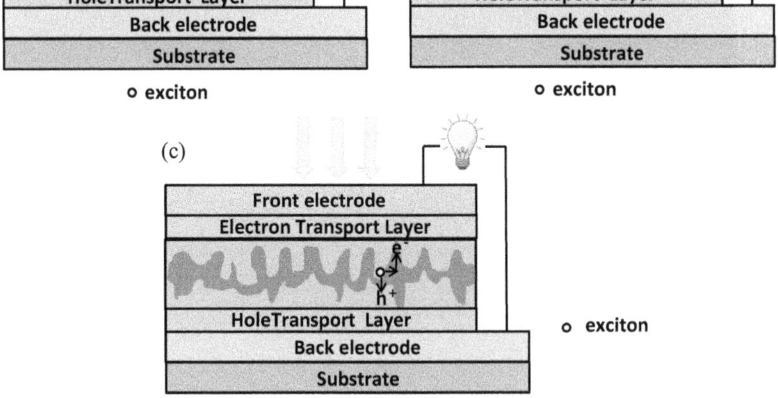

FIGURE 2.7 Working of e-donor/e-acceptor organic solar cells (a) single layer; (b) bilayer and (c) bulk heterojunction.

exciton excited in the acceptor reaches the donor-acceptor interface, the hole is transferred to the donor's HOMO. The dissociated electron is transported through the LUMO of the acceptor to the cathode. The dissociated hole is transported through the HOMO of the donor to the anode. The necessary condition herein is that the acceptor material has a LUMO that is lower than that of the donor material [19–21]. Figure 2.8 illuminates the energy band diagram of a bilayer solar cell.

The main limitation of bilayer organic solar cells is the exciton diffusion length. Only those excitons dissociate at the donor-acceptor interface, which are generated near the interface. So, to overcome this limitation, the concept of bulk heterojunction organic devices was proposed. The junctions in bulk heterojunction devices are formed by preparing a solution of two components and then casting this solution on an appropriate substrate. The two parts are then subjected to annealing, which enables them to self-assemble into a network that interconnects the two electrodes [22]. The nanometer size domain of this blend allows the excitons with short lifetimes to reach an interface and dissociate owing to the large donor–acceptor interfacial area (Figure 2.7c). These cells follow the same working mechanism as in bilayer organic devices as shown in Figure 2.8. Generally, Fullerenes such as C_{60} and its derivatives such as$PC_{71}BM$([6,6]-Phenyl-C_{71}-butyric acid methyl ester) and perylene are employed as electron acceptor layers. poly(3,4-ethylenedioxythiophene) & MEH-PPV (Poly[2-m ethoxy-5-(2′-ethylhexyloxy)-1,4-phenylene vinylene]), P3HT (poly(3-hexylthiophene)) and PCBM ([6,6]-phenyl-C_{61}-butyric acid methyl ester) are commonly used as electron donor layers. Polystyrenesulphonate (PEDOT:PSS), NiO, V_2O_5, and MoO_3 are employed as hole transport layers. Electron transport layers are made up of salts such as LiF, Cs_2CO_3, or organic compounds as bathophenantroline and bathocuproine.

2.3.3 Hybrid Junction-Based Cells

In the case of hybrid junction solar cells, both organic and inorganic compounds are utilized. In n-type dye-sensitized solar cells, the electrode is fabricated by depositing a thin layer of semiconducting oxides, such as TiO_2, Nb_2O_5, ZnO, or SnO_2, on a transparent conducting glass plate made of indium tin oxide (ITO). These oxides have a wide energy band gap ranging from 3 to 3.2 eV. Due to this wide band gap, these semiconducting materials absorb a very small amount of solar energy.

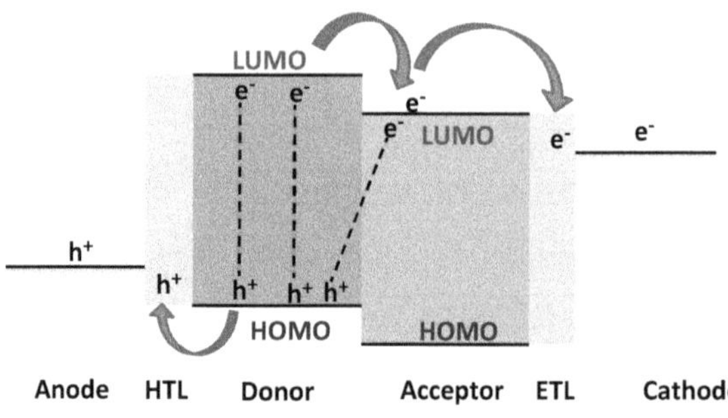

FIGURE 2.8 Schematic energy band diagram of donor-acceptor junction set up in bilayer cells.

These semiconducting films are soaked in the dye solution (such as ruthenium complex or various organic metal-free compounds) to lower the band gap, and a large number of dye molecules get covalently bonded to their surface. The cathode is crafted using Platinum or carbon. Both the electrodes are sealed together, and an electrolyte is filled with the help of a syringe. I^-/I_3^- liquid electrolyte is most commonly used in these cells. When sunlight passes through the electrode into the dye layer, dye molecules absorb photons of 1.72 eV energy and excite from their ground state (S) to a higher energy state (S*), as shown in Figure 2.9 [23,24].

The excited dye molecule (S*) is oxidized and an electron is injected into the conduction band of the semiconductor. Electrons are then transported to the anode. The oxidized dye molecule is again regenerated by electron donation from the iodide (I^-) in the electrolyte. In return, iodide (I^-) is regenerated by the reduction of triiodide (I_3^-) on the cathode. Similarly, in a p-type cell, the dye-sensitized semiconductor is of a p-type nature (typically nickel oxide). In a p-type semiconductor, instead of injecting an electron, a hole flows from the dye into the valence band of the semiconductor [23,24].

Many other hybrid solar cells have also been reported. These include π-conjugated polymers such as Poly(3-hexylthiophene) and poly (phenylene vinylene) as electron donors. The electron acceptor layer may contain CdSe quantum dots, PbS quantum dots [25], CdS, CdTe, TiO_2 nanoparticles [26], ZnO nanoparticles or graphene. The two materials are either assembled as bilayer or as bulk heterojunction devices, similar to organic solar cells as shown in Figure 2.7b and c [8,27–29]. The top electrode is made of TCO-coated glass, and the back electrode is made of Al/Pt/C in these hybrid cells.

Perovskite solar cells serve as another example of a hybrid junction type of solar cells. These cells use a perovskite-structured compound, most commonly a hybrid organic–inorganic lead or tin halide-based material. Perovskite is a compound having an ABX_3 crystal structure, where X represents halogens such as I^-, Br^-, Cl^- or oxygen, whereas A and B are cations of different sizes having 12 and 6 coordinates with X anions. The device structure contains three layers: a perovskite absorber layer, an electron transport layer, and a hole transport layer, as shown in Figure 2.10.

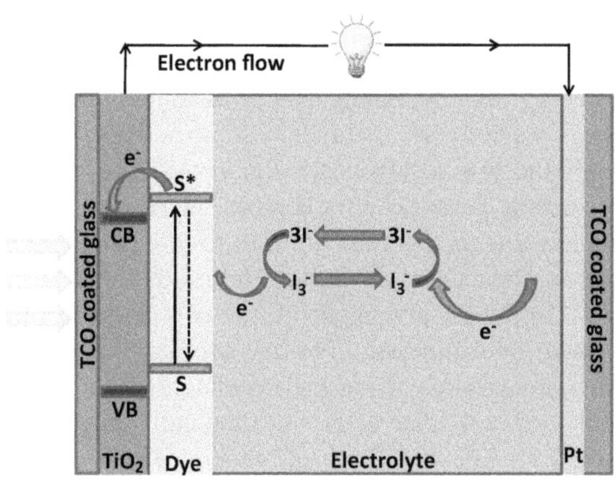

FIGURE 2.9 Schematic representation of working of dye sensitized solar cell.

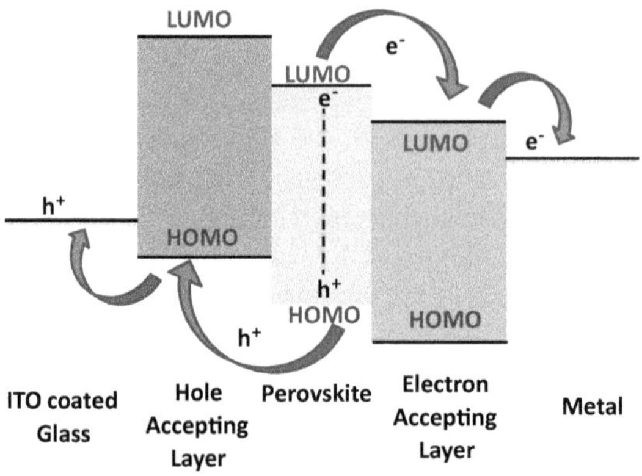

FIGURE 2.10 Schematic energy band diagram of Perovskite-based solar cells.

The perovskite layer, upon absorption of the incident radiation, generates electron and hole pairs. These generated electrons and holes are transported through the electron and hole transport layers, respectively, toward the external circuit with the help of charge-collecting contacts. FTO-coated glass is used as the top electrode, and the back electrode is a thermally evaporated gold layer [20]. The electron transport layer may consist of TiO_2, ZnO, CdS, fullerene derivative [6,6]-phenyl-C_{61}-butyric acid methyl ester (PCBM), or SnO_2. However, the hole transport layer can be comprised of organic materials such as poly(3,4-ethylenedioxythiophene) & polystyrene sulfonate (PEDOT:PSS), Spiro-OMeTAD (2,2′,7,7′-Tetrakis(*N*,*N*-di-*p*-methoxyphenylamine)9,9′-spirobifluorene) or inorganic materials like graphene oxide (GO), carbon (C), copper thiocyanate (CuSCN), copper zinc tin sulfide ($CuZnSnS_2$), copper indium disulfide ($CuInS_2$), copper iodide (CuI), cuprous oxide (Cu_2O), cupric oxide (CuO), nickel oxide (NiO) and various other materials [30,31].

There are two kinds of photons falling on the solar cells, one with lower energy than bandgap and one with higher energy than bandgap. The main problem with single-layer solar cells is that they do not absorb both kinds of photons. The photons with higher energy than the bandgap is utilized, but those with lower energy than the bandgap is wasted. So, in tandem solar cells, two kinds of photovoltaic cells with high and low bandgap are connected. The photovoltaic cell with a high band gap, such as perovskite, is kept on top so that high energy photons are absorbed, and low energy photons are sent to low band gap material such as Si. There are two approaches to designing these cells: two terminals (2T) and four terminals (4T). In the 2T arrangement, the two cells are joined in series by a recombination layer or a junction. The recombinant layer materials used are TCO, n- and p-doped nanocrystalline hydrogenated silicon (nc-Si:H) as well as nc-SiO*x*:H. In four terminal solar cells, both the cells are mechanically stacked as shown in Figure 2.11. The advantage of tandem cells is that their efficiency is high, and if one of the two cells stops working, the system will continue to function [32–34].

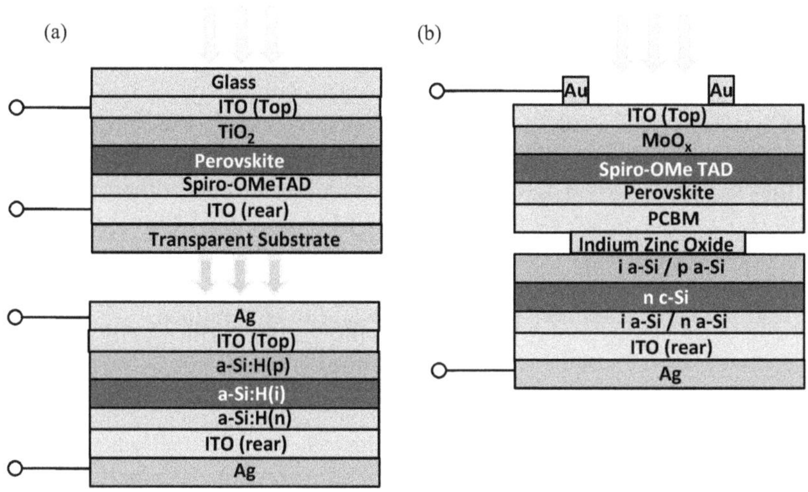

FIGURE 2.11 Schematic diagram of Tandem solar cells (a) four terminals and (b) two terminals.

2.4 SOLAR CELL PARAMETERS

The solar cells produce direct current electricity and the number of photons hitting the cell controls the current (*I*) flowing through the cell while the increase in temperature of the solar cell reduces its voltage. A simplified circuit diagram for a single-diode model of a solar cell is shown in Figure 2.12.

For this type of model, the *I–V* characteristics [35–37] can be expressed by the relation as

$$I = I_{\text{ph}} - I_O\left[\exp\left(\frac{V + IR_S}{nV_{\text{th}}}\right) - 1\right] - \frac{V + IR_S}{R_{\text{sh}}} \tag{2.2}$$

Where I is the output current of the cell, I_{ph} is photo-generated current, I_O is the saturation current of the equivalent diode, V is the output voltage of the cell, R_S and R_{sh} are the series and shunt resistances, n is the ideality factor of the cell >1, V_{th} is the thermal voltage of a cell and is given by

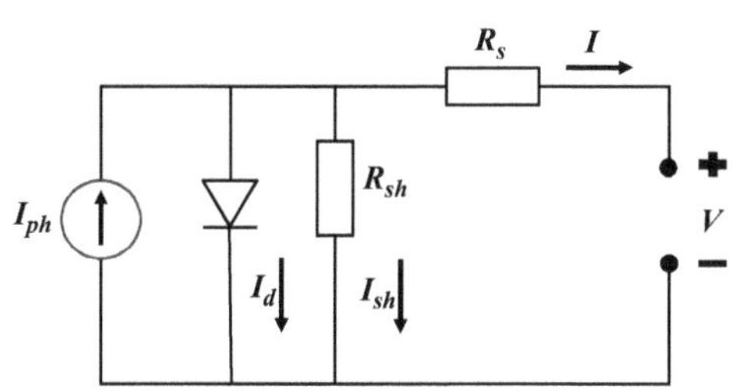

FIGURE 2.12 The simplified circuit diagram of a solar cell.

$$V_{th}(T) = \frac{kT}{q} \tag{2.3}$$

Where k is Boltzmann's constant, T is cell temperature, and q is electron charge. To evaluate the performance of the solar cell, various parameters are evaluated to define the efficiency of the cell. These cell parameters are embodied in the cell's current-voltage (I–V) values which are graphically plotted as shown in Figure 2.13 [35–37].

From this curve, various parameters such as short circuit current (I_{SC}), open circuit voltage (V_{OC}), maximum power point (P_M), current at maximum power point (I_{MP}), and voltage at maximum power point (V_{MP}) can be identified.

2.4.1 Short Circuit Current (I_{SC})

Short circuit current is the current flowing through the solar cell when the terminals of the cell are short-circuited (i.e. the voltage across the terminals of the solar cell is zero). This current is due to the generation and collection of light-generated electron-hole pairs. For an ideal cell, the light-generated current and short-circuited current are equal. Therefore, the short circuit current is the largest current that can be drawn from the solar cell. From Figure 2.13, I_{SC} is the maximum value of current represented along the current axis (y-axis).

$$I(\text{at } V = 0) = I_{SC} \tag{2.4}$$

2.4.2 Open Circuit Voltage (V_{OC})

Open circuit voltage (V_{OC}) is the voltage across the solar cell when the terminals of the cell are open (i.e. when the ideal load resistance across the cell is infinite). This is the maximum voltage that can be drawn across the cell when no current flows through the circuit. From Figure 2.13, V_{OC} is the maximum value of voltage represented along the voltage axis (x-axis).

$$V(\text{at } I = 0) = V_{OC} \tag{2.5}$$

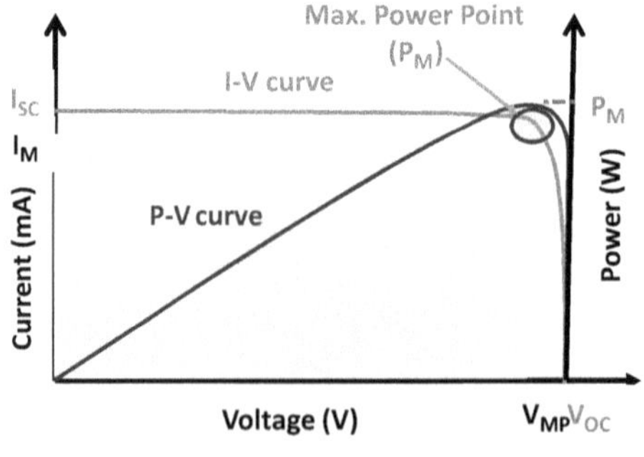

FIGURE 2.13 *I–V* and *P–V* characteristics curves of a solar cell.

2.4.3 Maximum Power Point (P_M)

The power produced by the cell is calculated by the product of the voltage and current drawn from the cell. At I_{SC} or V_{OC} points, the power drawn from the cell is zero. However, in the *I–V* curve, there is a point for which the product of the voltage (V_{MP}) and current (I_{MP}) is maximum, i.e. the power reaches its maximum value. This point is known as the maximum power point. In other words, it is the point where the cell generates maximum electrical energy. For ideal operations, the cell is defined at its maximum power point. From Figure 2.10, P_M is located near the bend of the *I–V* characteristics curve. The values of V_{MP} and I_{MP} can be estimated from the V_{OC} and I_{SC} of the solar cell [35,38]. The estimated values are $V_{MP} \approx (0.8–0.9)\ V_{OC}$ and $I_{MP} \approx (0.85–0.95)\ I_{SC}$. Also, the current and voltage of the cell depend upon the operating temperature, therefore, the estimated values of V_{MP} and I_{MP} depend upon the ambient temperature conditions.

$$P_M = V_{MP} \cdot I_{MP} \tag{2.6}$$

2.4.4 Fill Factor (FF)

The fill factor indicates the quality of the cell and internal energy losses in the cell. It relates the maximum power that a cell can provide in operating conditions and the product of the V_{OC} and I_{SC}. Mathematically,

$$\text{FF} = \frac{V_{MP} \cdot I_{MP}}{V_{OC} \cdot I_{SC}} \tag{2.7}$$

From Figure 2.13, FF visually represents how much the *I–V* curve deviates from a rectangular shape [39]. FF gives an idea of the quality of solar cells and its value lies close to unity. Typical values of FF are between 0.7 and 0.8.

2.4.5 Efficiency (η)

It is the ratio of the maximum electrical power output (P_{OUT}) that the cell can deliver to the solar power input (P_{in}) provided to the solar cell. The efficiency of the cell normally depends upon the cell fabrication type. Mathematically, it is given by

$$\eta = \frac{P_M}{P_{in}} = \frac{V_{MP} \cdot I_{MP}}{P_{in}} \tag{2.8}$$

Where V_{MP} is maximum point voltage, I_{MP} is the maximum point current, and the P_{in} is the solar power input.

P_{in} is found by taking the product of the irradiance of the light (in W/m^2) with the surface area of the cell (m^2). The efficiency of the cell also depends on the ambient conditions such as temperature and spectrum of incident light [39].

2.5 FACTORS AFFECTING THE POWER GENERATED BY SOLAR CELL

The solar cell is conventionally a p-n junction that works on the principle of photovoltaic effect. It is well known that sunlight is mainly converted into electrical energy. However, not all of the sunlight that reaches the surface of solar cells is converted into an electrical signal. It means loss of light is occurring, which affects the overall efficiency of solar cells. In the last few years, the efficiency of solar panel power generation has increased significantly which attracts many industries for investment in this area. The power generation with cost-effective ways of solar cells depends on many factors which are discussed as follows:

2.5.1 Conversion Efficiency (η)

The conversion efficiency of solar cells is one of the important parameters when we compare one solar cell with another. It is defined as the ratio of output from solar cells to input from the sun. The efficiency of solar cell is calculated by the following formula:

$$\eta = \frac{P_{\text{max}}}{E * A_c} \tag{2.9}$$

where P_{max} = maximum output power (W), E = Incident radiation flux (W/m^2), A = area of collector (m^2) [39]. The efficiency mainly depends on the intensity/spectrum of incident sunlight, environment conditions, reflection, recombination rate, and temperature. However, environmental conditions such as pollution, shading, and dust also play a very important role in the decrement of solar cell efficiency [40]. Therefore, it is important to consider the aforementioned factors as well as intrinsic and intermediate factors to optimize the performance of the solar cell. It is well known that, depending on the climate or seasons, the efficiency of solar cells are different. There is a big fallacy in our mind that temperature increases the efficiency of solar cells, but solar cells work under specific temperature; otherwise, high temperatures affect the overall efficiency of solar cells [39,40]. Furthermore, regular maintenance and monitoring of solar cells/panels again play an essential role in ensuring excellent performance and maximum solar cell efficiency. The parameters mainly affect the conversion efficiency of the solar cells is shown in Figure 2.14.

2.5.2 Amount of Input Light

Among many factors, the light intensity for the power generation performance of solar cells plays a very important role. The sunlight that reaches the earth's surface has wide wavelengths from ultraviolet to infrared region. When light falls on the surface of a solar cell, some of the photons are reflected, and others pass through it. The absorbed photons have a tendency to separate the electrons from atomic levels that create charge carriers and, hence, the electric current produced. It is well known that the current produced by solar cells is directly proportional to the amount of sunlight. However, the energy of some photons changes into heat, especially in traditional

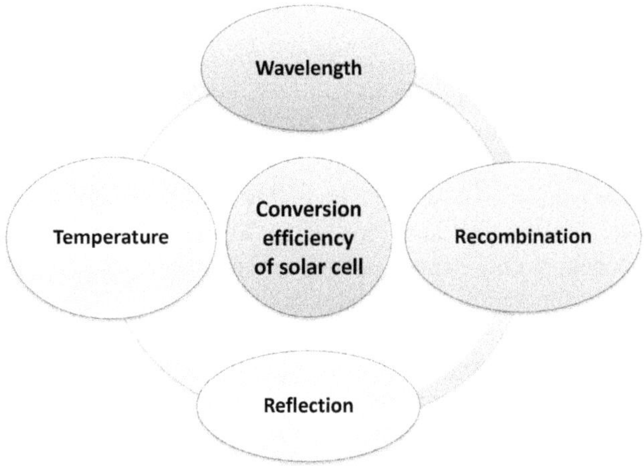

FIGURE 2.14 Parameters that affect conversion efficiency of solar cells.

solar cells. In traditional solar cells, the loss of energy is high as compared to the power generation, and the cost is very high. The new generation solar cell has less cost with good power generation using maximum input intensity [41].

The charge separation in an illuminated solar cell helps in the creation of potential differences. If this solar cell is further connected to the load, then the movement of the current that intiates the external circuit is positive and then returns to the cell through the negative. In this situation, solar cells behave as energy generators. In simple words, incident photons, when they fall on the cell with energy equal to or greater than the bandgap width, are quickly absorbed in the material and produce electron-hole pairs that initiate the current carrier in the solar cell [40,41].

2.5.3 Incident Angle of Light (θ) and Area of Solar Cell

The angle of incidence of light that reaches the solar cell/panel significantly affects the efficiency of the solar cell. The incident angle varies with the change of the location, time of year, and time of day. For instance, if the position of the solar cell/panel is in the south direction for those who are in the Northern hemisphere and the sun is also in the southern part of the sky, then solar captures the maximum amount of sunlight [42]. In addition, when sunlight strikes the earth at a perpendicular angle, then this produces the most intense impact. However, in most cases, the angle of incidence is lower; it means the intensity of sunlight is reduced by covering a large surface area of the solar panel. For example, if angle of incidence is 45° then the intensity of radiation is 30% less than 90°, because of the dispersion of light over the surface of the solar panel [43]. Therefore, the angle of radiation should be maintained according to the season change, sunshine hours, and longitude/latitude. In a similar way, the area of solar cells affects their efficiency such as larger cell/modules mean low efficiency and vice versa.

2.5.4 Operating Temperature (*T*)

The effect of temperature on the power efficiency of solar cells is one of the important parameters in solar technology because it is an intrinsic property of semiconductors. It is well known that, as the temperature increases, the efficiency of solar cells suddenly decreases and increases as solar cell temperature drops. Nowadays, commercial solar panels are designed with highly durable materials so that they can work in intense summer heat and cool winter weather. The temperature of commercial solar panels mainly depends on many factors such as geographic location, level of sunlight, temperature of air, and material of roof [44]. However, solar cells are usually made of crystalline Si, as Si is a good conductor of heat; it promotes heat in the solar cell on sunny days. Therefore, the effect of temperature on solar cell/panel output is important to calculate. It helps us to notify the performance of solar cells on summer sunny days. This can be done by using the temperature coefficient, which is generally mentioned in manufacturer's data sheet during the purchase. The temperature coefficient indicates a drop in panel efficiency when the temperature rises by 1°C; for instance, if the temperature coefficient if −0.5%, it means that the decrement in efficiency by 0.5% for each degree above 25°C and vice versa [44,45].

The two parameters such as V_{OC} and I_{SC} are mostly affected by an increase in the temperature in the solar cells. It is observed that the temperature above the optimal value decreases the V_{OC} and, ultimately, a low value of power as output is received and vice-versa [45]. The V_{OC} significantly decreases as temperature decreases due to temperature dependency on I_0.

$$I_0 = qA\frac{Dn_i^2}{LN_D} \tag{2.10}$$

In which q is a charge, A is an area, D is the diffusivity of the minority carrier, L is diffusion length for the minority carrier, N_D is a type of doping and n_i is intrinsic carrier concentration. In the aforementioned parameters, the intrinsic carrier concentration is one of the important temperature- dependent parameters [46].

The increase in I_{SC} is very slight as the temperature is increased, and hence, photons have sufficient energy for creating electron-hole pairs [42]. This small change in I_{SC} in Si solar can be defined as the following equation [47,48]:

$$\frac{1}{I_{SC}}\frac{dI_{SC}}{dT} \sim 0.0006 \text{ per } ^\circ\text{C } f \text{ for Si} \tag{2.11}$$

In Si, it can be 0.06% per °C.

The slight change in I_{SC} with temperature is mainly cell design dependent rather than properties of the material. In addition, the change in I_{SC} with temperature is always smaller than V_{OC}.

2.6 IDEAL OUTPUT CHARACTERISTICS OF SOLAR CELLS

The power output of a module depends on the number of cells in the module, the type of cells, and the total surface area of the cells. The output of a module depends on the amount of solar radiation, the angle of the module with respect to the sun, the temperature of the module, and the voltage at which the load (or battery) is drawing power from the module. Each solar cell and module has its own particular set of operating characteristics. These properties can be described by the current-voltage curve, which is better known as the *I–V* curve. *I–V* curves are used to compare various solar cell modules and to determine their performance at various levels of insulation and temperatures.

An ideal solar cell is defined by a current source in parallel with a single diode, as shown in Figure 2.15 [49].

If I, $I_{ph,}$ and I_0 are the output current, photocurrent, and reverse saturation current of the diode respectively for an ideal solar cell of known area A and ideality factor $\eta = 1$, the *I–V* characteristics in Equation 2.2 simplifies to the following equation:

$$I = I_{ph} - I_0\left[\exp\left(\frac{qV}{kT}\right) - 1\right] \tag{2.12}$$

where q is the electron charge, k is the Boltzmann's constant, T is the cell temperature, and V is the output voltage. The *I–V* curve (Figure 2.9) for a solar cell shows that the current is essentially constant over a range of output voltages for a specified amount of incident light energy [35,45]. Based on the *I–V* curve of a solar cell, the power–voltage (*P–V*) curve can be calculated as shown in Figure 2.13. It is advisable to operate a cell as near to the P_M as possible.

If a load is connected between the negative and positive sides of the cell, the electrons flow as a current. The power obtained from the ideal cell:

$$P = I^2 \cdot R_L = \left[I_{ph} - I_0\left(\exp\frac{qV_L}{kT} - 1\right)\right]^2 R_L \tag{2.13}$$

and the maximum power is obtained if R_L is properly adjusted and $dP_L/dR_L = 0$.

The maximum amount of energy (in watt-hours per day) that a solar cell produces per day can be roughly estimated by multiplying the maximum power of the cell (in W) by the number of peak sun hours received by the cell over the course of the day. The actual energy output will be less, depending on how near to the maximum power point the cell is operated.

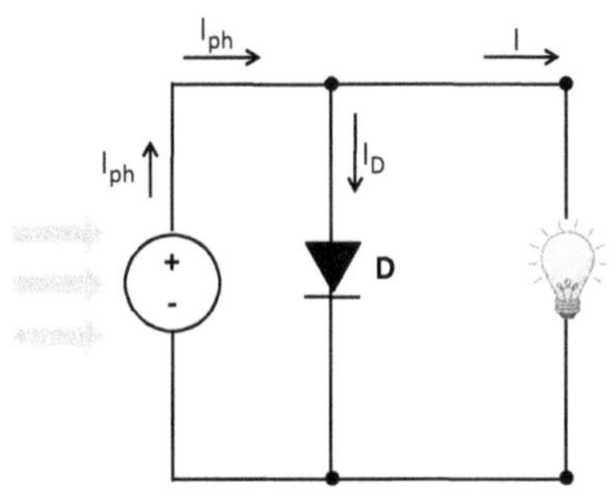

FIGURE 2.15 The simplified circuit diagram of an ideal solar cell.

2.7 CONCLUSION

Solar energy is a renewable source available as long as the sun exists, reducing reliance on non-renewable fuels. Utilizing solar energy decreases pollution, making it a clean and sustainable energy option. With the right strategies, solar power possesses the potential to be the largest energy source. Solar cells have come a long way in recent years due to technological advancements and increased industrial production and demand, making them a viable and cost-effective source of energy. With the declining cost of solar cells, we still have the potential to tap into even more solar resources to meet our energy needs. Switching from conventional energy sources like coal, gas, and oil to renewable energy sources is indeed crucial for mitigating the negative impacts on the environment, society, and the economy.

NOTE

* Corresponding author: Kapil Sood, Department of Physics, Government College Dhaliara, Himachal Pradesh, India.

REFERENCES

1. L.M. Fraas. History of solar cell development. In: *Low-Cost Solar Electric Power.* 2014, Springer.
2. K. Ranabhat, L. Patrikeev, A. A. Revina, K. Andrianov, V. Lapshinsky, E. Sofronova. An introduction to solar cell technology. *Journal of Applied Engineering Science* 2016, 405, 481–491.
3. Y. A. Jieb, E. Hossain. Solar system components. In: *Photovoltaic Systems.* 2022, 95–192. Springer.
4. S. G. Kumar, A. P. Shetty, C. R. Prashanth. Solar Cell Material Based on the Optimum Values of Key Parameters Using PC1D. In *Proceedings of the 2021 2nd International Conference for Emerging Technology (INCET)*, Belagavi, India, 21 May 2021; pp. 1–6.
5. A. Sarkin, N. Ekren, S. A. Saglam. Review of anti-reflection and self-cleaning coatings on photovoltaic panels. *Solar Energy* 2020, 199, 63–73.
6. H. K. Raut, V. A. Ganesh, A. S. Nairb, S. Ramakrishna. Anti-reflective coatings: A critical, in-depth review. *Energy and Environmental Science* 2011, 4, 3779–3804.
7. C. Ji, W. Liu, Y. Bao, X. Chen, G. Yang, B. Wei, F. Yang, X. Wang. Recent applications of antireflection coatings in solar cells. *Photonics* 2022, 9, 906.
8. J. A. Luceno-Sanchez, A. M. Díez-Pascual, C. R. Peña. Materials for photovoltaics: State of art and recent developments. *International Journal of Molecular Sciences* 2019, 20, 976.
9. F. Liu, W. Wang, L. Wang, G. Yang. Working principles of solar and other energy conversion cells. *Nanomaterials and Energy* 2013, 2, 3–10.
10. R. Swami. Solar cell. *International Journal of Scientific and Research Publications* 2012, 2, 1–5.
11. S. R. Wenham, M. A. Green. Silicon solar cells. *Progress in Photovoltaics* 1996, 4, 3–33.
12. A. De Vos. Thermodynamics of photochemical solar energy conversion. *Solar Energy Materials and Solar Cells* 1995, 38, 11–22.
13. R. M. Geisthardt, M. Topic, J. R. Sites. Status and potential of CdTe solar-cell efficiency. *IEEE Journal of Photovoltaics* 2015, 5, 1217–1221.

14. J. Britt, C. Ferekides. Thin-film CdS/CdTe solar cell with 15.8% efficiency. *Applied Physics Letters* 1993, 62, 2851–2852.
15. M. C. Scharber, N. S. Sariciftci. Low band gap conjugated semiconducting polymers. *Advance Materials Technologies* 2021, 6, 200857.
16. J. Ramanujam, U. P. Singh. Copper indium gallium selenide based solar cells – A review. *Energy & Environment Science* 2017, 10, 1306–1319.
17. M. P. Suryawanshi, G. L. Agawane, S. M. Bhosale, S. W. Shin, P. S. Patil, J. H. Kim, A. V. Moholkar. CZTS based thin film solar cells: A status review. *Materials Technology* 2013, 28, 98–109.
18. H. Lin, M. Yang, X. Ru, G. Wang, S. Yin, F. Peng, C. Hong, M. Qu, J. Lu, L. Fang, C. Han. Silicon heterojunction solar cells with up to 26.81% efficiency achieved by electrically optimized nanocrystalline-silicon hole contact layers. *Nature Energy* 2023, 8, 789–799.
19. J. Nelson. Organic photovoltaic films. *Current Opinion in Solid State and Materials Science* 2002, 6, 87–95.
20. N. Marinova, S. Valero, J. L. Delgado. Organic and perovskite solar cells: Working principles, materials and interfaces. *Journal of Colloid and Interface Science* 2017, 488, 373–389.
21. H. Hoppe, N. S. Sariciftci. Organic solar cells: An overview. *Journal of Materials Research* 2004, 19, 1924–1945.
22. C. Deibel, V. Dyakonov, C. J. Brabec. Organic bulk-heterojunction solar cells. *IEEE Journal of Selected Topics in Quantum Electronics* 2010, 16, 1517–1527.
23. K. Sharma, V. Sharma, S. S. Sharma. Dye-sensitized solar cells: Fundamentals and current status. *Nanoscale Research Letters* 2018, 13, 1–46.
24. M. K. Nazeeruddin, E. Baranoff, M. Grätzel. Dye-sensitized solar cells: A brief overview. *Solar Energy* 2011, 85, 1172–1178.
25. T. Blachowicz, A. Ehrmann. Recent developments of solar cells from PbS colloidal quantum dots. *Applied Sciences* 2020, 10, 1743.
26. T. F. Guo, G. L. Pakhomov, T. C. Wen, X. G. Chin. S. H. Liou. Effect of TiO_2 nanoparticles on polymer-based bulk heterojunction solar cells. *Japanese Journal of Applied Physics* 2006, 45, L1314.
27. M. Wright, A. Uddin. Organic—inorganic hybrid solar cells: A comparative review. *Solar Energy Materials and Solar Cells* 2012, 107, 87–111.
28. X. Fan, M. Zhang, X. Wang, F. Yang, X. Meng. Recent progress in organic–inorganic hybrid solar cells. *Journal of Materials Chemistry A* 2013, 1, 8694–8709.
29. P. V. Kamat, Quantum dot solar cells. The next big thing in photovoltaics. *The Journal of Physical Chemistry Letters* 2013, 4, 908–918.
30. C. Zuo, H. J. Bolink, H. Han, J. Huang, D. Cahen, L. Ding. Advances in perovskite solar cells. *Advanced Science* 2016, 3, 1500324.
31. N. G. Park. Perovskite solar cells: An emerging photovoltaic technology. *Materials Today* 2015, 18, 65–72.
32. P. Löper, S.-J. Moon, S. M. de Nicolas, B. Niesen, M. Ledinsky, S. Nicolay, J. Bailat, J.-H. Yum, S. D. Wolf, C. Ballif. Organic–inorganic halide perovskite/crystalline silicon four-terminal tandem solar cells. *Physical Chemistry Chemical Physics* 2015, 17, 1619–1629.
33. E. Raza, Z. Ahmad. Review on two-terminal and four-terminal crystalline-silicon/perovskite tandem solar cells; progress, challenges, and future perspectives. *Energy Reports* 2022, 8, 5820–5851.
34. N. Liu, L. Wang, F. Xu, J. Wu, T. Song, Q. Chen. Recent progress in developing monolithic perovskite/Si tandem solar cells. *Frontiers in Chemistry* 2020, 8, 1120.
35. L. Peng, Y. Sun, Z. Meng, Y. Wang, Y. Xu. A new method for determining the characteristics of solar cells. *Journal of Power Sources* 2013, 227, 131–136.

36. J. I. Rosell, M. Ibáñez. Modelling power output in photovoltaic modules for outdoor operating conditions. *Energy Conversion and Management* 2006, 47, 2424–2430.
37. T. Dittrich. Basic characteristics and characterization of solar cells. In *Materials Concepts for Solar Cells*, 2018, 3–43. World Scientific.
38. A. Q. Malik, S. J. B. H. Damit. Outdoor testing of single crystal silicon solar cells. *Renewable Energy* 2003, 28, 1433–1445.
39. B. H. Hamadani, B. Dougherty. Solar cell characterization. In M. P. Paranthaman, W. Wong-Ng, R. Bhattacharya (eds.), *Semiconductor Materials for Solar Photovoltaic Cells*, 2016, 218. Springer, Cham.
40. L. D. Jathar, S. Ganesan, U. Awasarmol, K. Nikam, K. Shahapurkar, M. E. M. Soudagar, H. Fayaz, A. S. El-Shafay, M. A. Kalam, S. Bouadila, S. Baddadi, V. Tirth, A. S. Nizami, S. S. Lam, M. Rehan. Comprehensive review of environmental factors influencing the performance of photovoltaic panels: Concern over emissions at various phases throughout the lifecycle. *Environmental Pollution* 2023, 326, 121474.
41. S. W. Yufenyuy, G. M. Mengata, L. N. Nneme, U. M. Bongwirnso. Indoor environment PV applications: Estimation of the maximum harvestable power. *Renewable and Sustainable Energy Reviews* 2024, 193, 114287.
42. T. Feurer, P. Reinhard, E. Avancini, B. Bissig, J. Löckinger, P. Fuchs, R. Carron, T. P. Weiss, J. Perrenoud, S. Stutterheim, S. Buecheler, A. N. Tiwari. Progress in thin film CIGS photovoltaics–Research and development, manufacturing, and applications. *Progress in Photovoltaics: Research and Applications* 2017, 25, 645–667.
43. R. Parthiban, P. Ponnambalam. An enhancement of the solar panel efficiency: A comprehensive review. *Frontiers in Energy Research* 2022, 10, 937155.
44. S. Dubey, J. N. Sarvaiya, B. Seshadri. Temperature dependent photovoltaic (PV) efficiency and its effect on PV production in the world – A review. *Energy Procedia* 2013, 33, 311–321.
45. M. Sani, A. Sule. Effect of temperature on the performance of photovoltaic module. *International Journal of Innovative Science and Research Technology* 2020, 5, 670–676.
46. C. Sun, Y. Zou, C. Qin, B. Zhang, X. Wu. Temperature effect of photovoltaic cells: A review. *Advanced Composites and Hybrid Materials* 2022, 5, 2675–2699.
47. S. Pizzini, C. Calligarich. On the effect of impurities on the photovoltaic behavior of solar-grade silicon. *Journal of the Electrochemical Society* 1984, 131, 2128.
48. P. J. Cousins, D. D. Smith, H.-C. Luan, J. Manning, T. D. Dennis, A. Waldhauer. Gen III: Improved Performance at Lower Cost. In *35th IEEE Photovoltaic Specialists Conference*, Honolulu, Hawaii, 2010.
49. W. Ali, P. Cofie, J. Fuller, S. Lokesh, E. Kolawole. Performance and efficiency simulation study of a smart-grid connected photovoltaic system. *Energy and Power Engineering* 2017, 9, 71–85.

3 Fabrication of Silicon-Based Solar Cell

*Barkha Tyagi**

3.1 INTRODUCTION TO SI-BASED FABRICATION TECHNOLOGY

3.1.1 Introduction to Si Wafer

In electronics, a wafer is a thin slice of semiconductor such as crystalline silicon which serves as a substrate for fabrication of electronic devices such as photovoltaic (PV). Silicon wafers play a pivotal role in the technology sector, serving as the foundational material for microelectronic devices. These thin films of semiconductor material are essential for the creation of integrated circuits (ICs), which find applications in various electronic devices such as smartphones, computers, medical equipment, and aerospace technology. Silicon, a vital semiconductor, is popular in the electronic and technological industries owing to its conductivity and cost-effectiveness. It stands as the seventh most abundant element universally and the second most abundant on Earth, but it is not found in its pure form. Everyday materials like beach sand, quartz, and flint are some common sources of silicon [1].

Pure silicon functions as an insulator, permitting minimal electricity flow, yet this characteristic undergoes alteration through a process known as doping. Doping involves incorporating a small amount of impurity into a silicon crystal to modify its properties and render it conductive. Doping will be discussed in more detail later in this chapter. These substances used for doping are referred to as dopants. Silicon, in its pure form, exhibits poor electrical conductivity; however, it can be doped precisely to regulate resistivity according to specific requirements. Resistivity, measured in ohm-centimeters (ohm-cm or Ω·cm), quantifies the material's ability to impede the flow of electrical current. Dopants such as nitrogen, indium, aluminum, gallium, and boron can be introduced during the silicon growth process. Consequently, to transform non-conductive silicon into a semiconductor, silicon must be processed into wafers, thus yielding a silicon wafer, which is available in various shapes and sizes.

Manufacturing silicon wafers involves various steps such as extracting the pure silicon, purification, fabricating cylindrical ingots (boule) of silicon, wafer slicing, polishing, and cleaning. After extracting raw silicon, it undergoes rigorous purification to eliminate impurities, crucial for electronic device performance. This purification process, involves refining, zone melting, and solidification, which increases silicon purity to 99.9999% or higher. Common fabrication methods like the Floating Zone or Czochralski process are used to solidify silicon into rods or ingots. The ingots are then sliced into disc-shaped wafers of uniform thickness using diamond saw blades. Typically, around 1 mm thick, wafers can be thinner for specific applications.

DOI: 10.1201/9781003499701-3

Subsequently, wafers are polished to ensure a smooth, flat surface suitable for electronic device fabrication. The Czochralski process, named after Jan Czochralski, produces single crystals of semiconductors, metals, salts, and synthetic gemstones. Initially developed in 1916, it's chiefly employed for fabricating single-crystal silicon crucial for semiconductor devices. It entails dipping a seed crystal into molten silicon within a crucible, then slowly pulling and rotating the seed crystal upwards. This controlled process ensures uniform crystal structure, resulting in a large cylindrical silicon crystal, or boule. Sliced into wafers, these crystals undergo polishing for use in semiconductor manufacturing. Renowned for its ability to produce high-quality, defect-free crystals, the Czochralski process is preferred in the semiconductor industry.

Following the creation of the silicon boule via the Czochralski process, the subsequent stage in silicon wafer production involves slicing and polishing. The boule, a cylindrical single crystal, is cut into thin discs or wafers using a diamond saw, known for its precision due to its hardness. Wafer thickness varies based on semiconductor device requirements but typically is around 1 mm. Precision slicing is crucial to ensure uniform thickness and defect-free wafers, vital for semiconductor device performance. Hence, the slicing process is pivotal in silicon wafer production.

Silicon wafers offer a distinctive blend of physical and chemical attributes, rendering them crucial in the technology sector. These include electrical and thermal conductivity, as well as mechanical robustness. Comprehending these traits is crucial for crafting reliable electronic devices, as they significantly influence performance and durability.

Silicon wafers possess crucial electrical conductivity, being semiconductors with a conductivity range between conductors like copper and insulators like glass. Doping, involving small impurity additions, precisely tunes these properties. Electron-donating or accepting dopants like phosphorus, arsenic, boron, or aluminum are used, generating n-type or p-type silicon by altering electron concentration. This controlled doping enables tailored electrical properties crucial for semiconductor device fabrication. The precise control of electrical properties in silicon wafers is essential for developing electronic components such as transistors, diodes, and integrated circuits. These devices rely on regulated current flow, achievable through silicon's unique electrical characteristics. The quality and consistency of silicon wafers directly impact the performance, efficiency, and reliability of electronic devices.

Silicon wafers possess vital thermal properties crucial for electronic device functionality. Thermal conductivity, measuring a material's heat conduction ability, is high for silicon, approximately 149 W/m·K at room temperature. This property efficiently dissipates heat during device operation, ensuring reliability and longevity by preventing overheating.

Another critical property is the coefficient of thermal expansion (CTE), indicating material expansion or contraction with temperature changes. At room temperature, silicon exhibits a relatively low CTE, about 2.6 ppm/K. This stable behavior prevents significant dimensional changes during temperature shifts, averting mechanical stresses and potential device failure. These thermal properties are essential not only for the fabrication process of electronic devices but also for device operation for applications in PV.

Silicon wafers are pivotal in solar cell production and fundamental to solar panels for harvesting solar energy. Solar cells, or PV cells, convert sunlight into electricity through the PV effect. Silicon's exceptional semiconductor properties, alongside its sunlight absorption capacity, render it ideal for solar cells. When sunlight interacts with silicon wafers, it energizes electrons, generating electric current.

Solar cells primarily employ two types of silicon: monocrystalline and polycrystalline. Monocrystalline silicon, with a single crystal structure, ensures efficient electron flow, leading to high efficiency. Polycrystalline silicon, comprising multiple crystal structures, may hinder electron flow, resulting in lower efficiency but it is cost-effective.

The production of silicon wafers for solar cells involves processes akin to semiconductor industry methods, including the Czochralski process, wafer slicing, and polishing. Despite variances in thickness and purity compared to semiconductor wafers, silicon wafers' fundamental electrical and thermal properties remain critical in solar cell fabrication.

3.1.2 Cleaning the Si Wafer

The evolution of silicon wafer cleaning technology can be traced back to the 1950s. For good device performance and reliability of ultra-large-scale integration (ULSI) silicon circuit technology, techniques to avoid contamination and processes to generate ultra-clean wafer surfaces have become critically important.

When we use the term "ultra-clean", it includes diminishing chemical contaminations and particles on the wafer surface. The majority of yield losses in integrated circuit fabrication are generally accepted to be due to micro-contamination.

Figure 3.1 shows some trace impurities, such as sodium ions, metals, and particles, which can be detrimental if present on semiconductor surfaces during high-temperature processing, such as thermal oxidation, diffusion, and epitaxial growth because they can diffuse into the interior of the semiconductor. Silicon wafers are susceptible to various types of contaminants, which can have adverse effects on the process, leading to the incorporation of pattern or ion implantation defects. The types of contaminants over silicon wafers are listed below in Table 3.1.

Moreover, the removal of impurities before and after lower temperature steps, such as chemical vapor deposition, dopant implanting, and plasma reaction, is also critical. The other term used is "surface conditioning," which refers to the fact that the process of cleaning can do more than just remove impurities. It can also prepare the surface chemically for the next process; for example, immersing the Si wafer into a diluted HF

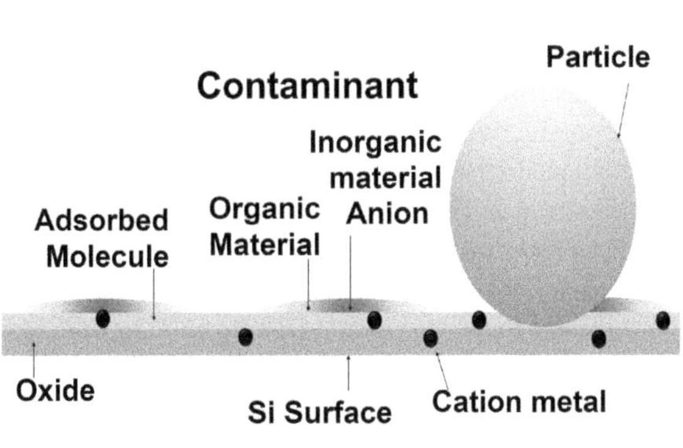

FIGURE 3.1 Surface contaminants that may be present on a silicon wafer.

TABLE 3.1
Types of Contaminants over Silicon Wafer and Their Effects on the Process

Type of Contamination		Effect on the Wafer/Device
Particle contamination		Shunt path in the device Loss in performance
Metallic contamination	Alkali metals	Breakdown/degradation
	Heavy metal	Loss of current Lifetime degradation
Chemical contamination	Inorganic dopants	Device voltage shifts Si substrate sheet resistance variations
	Organic materials	Film uniformity variations Haze occurrence
	Inorganic acids and bases	Haze due to salt generation Recombination losses
	Oxidation due to moisture, air	Increased contact resistance at the interface leads to degradation

solution can clean the surface from oxide species and leave a hydrogenated Si surface, which assists the epitaxial layer growth in the gas phase. The purpose of wafer cleaning is to remove particles and chemical impurities from the semiconductor surface without damaging the substrate surface, causing a change in roughness, pitting, or corroding. To achieve this purpose techniques like plasma, dry-physical, wet chemical, vapor phase, and supercritical fluid methods can be used [2].

3.2 FABRICATION OF DIODE

3.2.1 Concept of Doping

To understand doping, we need to understand intrinsic and extrinsic semiconductors. An intrinsic semiconductor is the pure form of any semiconductor (free of any impurities) where electron-hole pairs can be generated even at room temperature. Under the influence of an electric field, both electrons and holes are responsible for the conduction, and the total current inside the semiconductor is the sum of currents due to free electrons and holes. However, the current conduction inside intrinsic semiconductors is insufficient to be used in electronic devices. Therefore, it is necessary to alter the conductivity of pure semiconductors to increase their conductivity. This can be achieved through a technique known as "doping". The doping of a semiconductor material is the process of introducing impurities. The amount and type of such impurities must be closely controlled during the preparation of extrinsic semiconductors. The purpose of doping is to increase either the number of free electrons or holes in the semiconductor material. Depending on what type of impurity is added to the semiconductor, extrinsic semiconductors can be classified into two types, namely, n-type and p-type semiconductors, as shown in Figure 3.2.

To increase the number of free electrons in the semiconductor, a pentavalent impurity (having 5 valence electrons) is added and the semiconductor is called n-type.

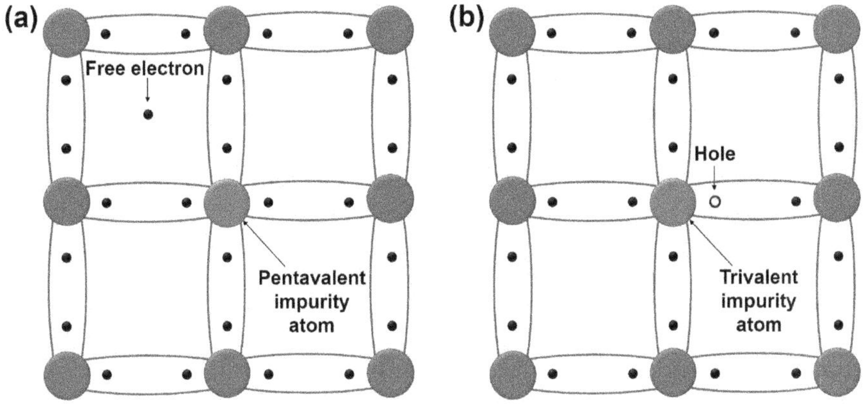

FIGURE 3.2 Doping of an intrinsic semiconductor to convert into n-type (a), or p-type (b) semiconductor.

To increase the number of free holes, a trivalent impurity (having 3 valence electrons) is added and the semiconductor is called p-type. Therefore, we end up with a larger number of free electrons in the n-type material and a larger number of holes in the p-type material. However, it is important to understand that some of the covalent bonds in these semiconductor materials might break even at room temperature, thus releasing an equal number of free electrons and holes apart from the incorporated impurity. Thus, in an n-type material, the larger number of free electrons are called the majority carriers, and the small number of holes are called the minority carriers, and in a p-type material the larger number of holes are called the majority carriers, and the small number of electrons are called the minority carriers. When these n-type and p-region semiconductors are joined, the contact surface is called a p-n junction, as shown in Figure 3.3. Most semiconductor devices contain one or more p-n junctions.

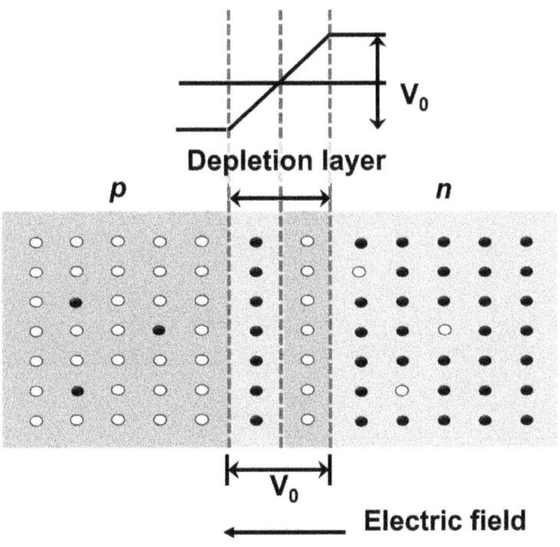

FIGURE 3.3 Schematic representation of a p-n junction.

During the formation of a p-n junction, the free electrons near the junction in the n region start to diffuse across the junction into the p-region where they combine with the holes near the junction. This leads to the n-region losing free electrons and creating a layer of positive charges near the junction. As the electrons move across the junction, the p-region loses holes as the electrons and holes combine, resulting in a layer of negative charges near the junction. These newly formed layers of positive and negative charges form the depletion region. This region is depleted of charge carriers due to diffusion across the junction. Once the p-n junction is formed and the depletion layer is created, there is no further diffusion of free electrons because the depletion region acts as a barrier to prevent the movement of free electrons across the junction. When this happens, the junction is in equilibrium. Moreover, due to the presence of positive and negative charges, an electric field comes into play, resulting in a potential difference (barrier potential (V_0)) across the depletion layer. The barrier potential for silicon diode is approximately 0.7 V.

3.2.2 Equilibrium of Fermi Energy Level

The valance band of a semiconductor is a denotation of the energy level of the electrons in the valence shell of the atomic structure. On the other hand, the conduction band is the delocalized band of energy levels in a semiconductor that are partially filled with electrons. The electrons are responsible for conduction. At absolute zero temperature, all electrons lie in the lowest energy state. The highest energy level that an electron can occupy at absolute zero temperature is called the fermi level and it lies between the valence band and the conduction band. All band energies are measured with respect to the vacuum level. The vacuum level represents the energy of a stationary electron outside the semiconductor. Electron affinity (symbolized by χ in eV) signifies the energy needed to transfer an electron from the bottom of the conduction band to the vacuum level, remaining constant for a material. In-band alignment, different semiconductor heterojunctions align their bands based on their electron affinities. This alignment dictates that the conduction band offset at the interface equals the difference in electron affinities between the semiconductors, as depicted in the accompanying Figure 3.4.

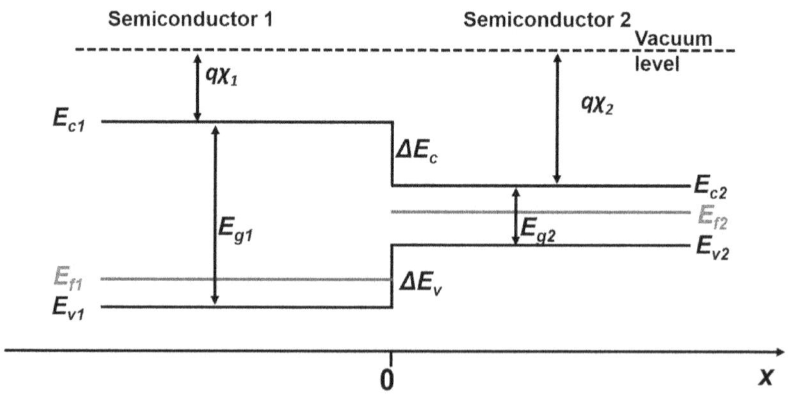

FIGURE 3.4 Band-diagram of a p-n heterojunction diode before attaining thermal equilibrium.

According to the electron affinity rule the conduction band offset ΔE_c is given as,

$$\Delta E_c = q(\chi_2 - \chi_1) \tag{3.1}$$

Then the valence band offset is given by,

$$\Delta E_v = (E_{g1} - E_{g2}) - \Delta E_c = \Delta E_g - \Delta E_c \tag{3.2}$$

Thus from Equation 3.1 and 3.2, we get,

$$\Delta E_c + \Delta E_v = \Delta E_g \tag{3.3}$$

Now, assuming that the conduction band offset ΔE_c is known, the doping in semiconductor 1 is p-type and equal to N_a, the doping in semiconductor 2 is n-type and equal to N_d, and all dopants are ionized. The resulting band alignment is shown in Figure 3.4; however, it is not a correct description at equilibrium as the Fermi level is not constant throughout the structure. There are more electrons in the region $x>0$ than in the region $x<0$, and more holes in the region $x<0$ than in the region $x>0$. As a consequence, electrons move from the n-side to the p-side while holes move from the p-side to the n-side. The presence of ionized donor atoms contributes electrons on the n-side. When electrons diffuse into the p-side, they leave positively charged donor atoms behind, and similarly, when holes diffuse into the n-side, they leave negatively charged acceptor atoms behind. This results in a net positive charge density on the right side ($x>0$) and a net negative charge density on the left side ($x<0$) of the interface, forming depletion regions. These regions lack the majority of charge carriers. The charge imbalance generates an electric field from donor to acceptor atoms as displayed in Figure 3.3. The electric field opposes the diffusive currents of electrons and holes, eventually reaching equilibrium when their drift and diffusive components balance out.

3.2.3 Band Bending

Band bending implies an electric field and, therefore, a potential difference across the junction. The band bending that occurs inside the depletion region reflects the presence of an electric field and a corresponding electrostatic potential. The vacuum level also bends in response to the electric field, as shown in Figure 3.5. The bands bend because the fermi level in equilibrium is flat and constant throughout the device.

$$qV_{bi} = E_{f2} - E_{f1} \tag{3.4}$$

Since,

$$E_{f2} - E_{c2} = KT \ln\left(\frac{N_d}{N_{c2}}\right) \tag{3.5}$$

$$E_{v1} - E_{f1} = KT \ln\left(\frac{N_a}{N_{v1}}\right) \tag{3.6}$$

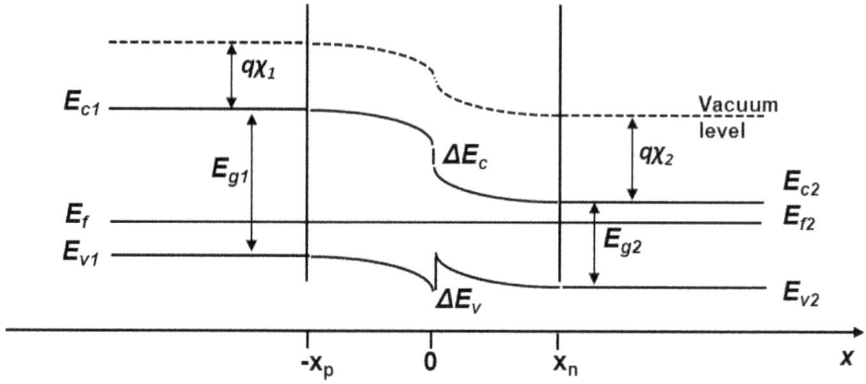

FIGURE 3.5 Band-diagram of a p-n heterojunction diode after attaining thermal equilibrium.

Summing the above equations gives,

$$qV_{\text{bi}} = E_{f2} - E_{f1} = E_{c2} - E_{v1} + KT\ln\left(\frac{N_a N_d}{N_{c2} N_{v1}}\right) \tag{3.7}$$

$$qV_{\text{bi}} = E_{g1} - \Delta E_c + KT\ln\left(\frac{N_a N_d}{N_{c2} N_{v1}}\right) \tag{3.8}$$

$$qV_{\text{bi}} = E_{g2} + \Delta E_v + KT\ln\left(\frac{N_a N_d}{N_{c2} N_{v1}}\right) \tag{3.9}$$

3.3 DESIGNING AND PATTERNING OF ELECTRICAL CONTACTS

Semiconductors like silicon are significantly less conductive than metals, necessitating a transparent conductor and grid pattern on top to diminish series resistance. In a solar cell, the electrical contacts are designed to admit light into the cell while conveying electricity outwards, which means the top contact should be transparent enough such as indium tin oxide. Transparent conductors such as indium tin oxide are less conductive than metal. Therefore, metallic top contacts are vital for collecting the current produced by a solar cell. However, the metal wires are opaque in nature and block light from reaching the silicon absorber layer; therefore, they should be thin. Thus, the design of the top contact involves a trade-off between shading the cell, which predominantly impacts short-circuit current density, and the resistance of the metal contact, which predominantly affects the fill factor. The metal wires of the top contact could be exceedingly narrow and closely spaced, but such fine contacts are prohibitively expensive for solar applications. For example, photolithography achieves line widths below 1 μm and is extensively used in the integrated circuit industry, but it requires costly chemicals and intricate alignment. Photolithography is seldom used in commercial cells due to its expense, typically being reserved for laboratory demonstration cells. Therefore, on an industrial scale screen printing is the most commonly used technique to print top silver electrodes. "Busbars" are directly linked to the external leads, while "fingers" are finer areas of metallization that gather current for delivery to the busbars. The primary design trade-off in top contact design revolves around balancing the increased resistive losses associated with a widely spaced grid and the amplified reflection caused by a high fraction of metal coverage on the top surface.

3.4 ROLE OF ANTI-REFLECTIVE COATING

The optical phenomenon of reflection occurs as a result of optical disturbance when light is traveling from one medium to another having different refractive indices, which is an optical parameter quantifying the speed of light in a given medium. Optical instruments and optoelectronic devices have an ever-growing demand for exploring new ways of maximizing light collection efficiency. Likewise, the performance of silicon solar cells is limited by the high reflection of over 30% at the surface of the cell. These reflection losses can be reduced by applying different strategies like anti-reflection coating (ARC), texturing the surface, etc., which will be discussed in this chapter. The technique of applying ARC over the surface of silicon solar cells is similar to those used in optics such as camera lenses. To reduce the reflection losses, a thin layer of dielectric material is applied on the surface. The whole anti-reflection effect is attained using interference effects in these thin layers of dielectric materials with a specifically chosen thickness as shown in Figure 3.6. Materials such as SiN_x, TiO_2, and Al_2O_3 are commonly used as ARCs in silicon solar cells.

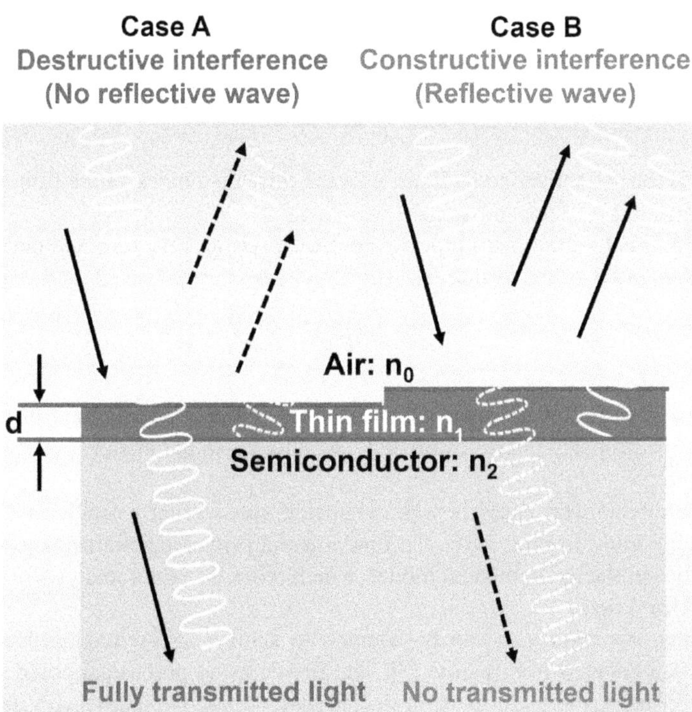

FIGURE 3.6 Constructive and destructive interference of light when passing through one medium to the other.

To understand the basic concept of ARCs, one needs to understand the Fresnel equation, which offers the basic preliminary mathematical model of reflection and refraction. Augustin-Jean Fresnel introduced reflectance (R) as the fraction of incident light reflected at the interface, while transmittance (T) measures the remaining light transmitted (refracted). The mathematical model, or vector method, used to derive conditions for anti-reflection involves considering the refractive index of a thin film and a glass substrate, based on certain assumptions

- Reflected waves have equal intensity and only one reflected wave per interface.
- Other optical interactions like scattering and absorption are negligible.

Thus, if the two reflected waves undergo destructive interference, canceling each other out, there would be no reflection. This leads to two essential criteria for Anti-Reflection:

- Reflected waves are π radians out of phase, or the phase difference (δ) is $n\pi/2$.
- The film thickness (d) is an odd multiple of $\lambda/4$, where λ is the wavelength of the incident beam.

Using the equation $\delta = 2\pi nd \cos\theta/\lambda$ for phase difference and substituting the values of δ and d, we find $\theta = 0$, indicating normal incidence. The reflectance at normal incidence is given by:

$$R = \left[\frac{n_0\, n_2 - n_1^2}{n_0\, n_2 + n_1^2} \right]^2 \tag{3.10}$$

Where n_0 is the refractive index of air, n_1 is the refractive index of the film, and n_2 is the refractive index of the substrate.

As the aim is to attain zero reflectance, hence setting R to zero and determining the refractive index of the film (n) as:

$$n_1 = \sqrt{n_0 n_2} \tag{3.11}$$

Since R as proposed by Fresnel, depends on the s and p-polarization of light, the anti-reflective property is also evaluated based on these parameters, as will be discussed later.

The aforementioned analysis was conducted considering a single layer on the substrate. For multi-layered ARC, the fundamental principle remains consistent; the variation lies in the mathematical model, which relies on vector analysis of the individual reflected rays.

Generally, two strategies can be adopted to achieve anti-reflection; basic models have been adopted from nature [3]. The first is using porous/patterned surfaces. Introducing such patterns to the film decreases its refractive index, with the dependence on volume fraction, as the nanostructured, sponge-like nature of patterned silicon allows for lowering the refractive index of silicon with respect to air. An example of this exists in nature as an absolutely lusterless corneal lens of the eye of the moth that helps track predators. A Moth's eye consists of a nanostructure pattern having dimensions smaller than the wavelength of light, which helps prevent light reflect from it. Generally, when light falls on macroscopic structures, it will undergo reflection and scattering after being partly absorbed. However, when light falls on nanostructures, it interacts with the structure as the wavelength of light, which is larger than the dimension of the structure, as if the surface possesses a gradient, causing the light waves to bend gradually. Moreover, the spacing and depth of the nanostructures can be tuned to trap the light waves causing multiple internal reflections. This leads to the absorption of the maximum amount of the incident radiation, as shown in Figure 3.7.

Graded refractive index (GRIN) is another way of reducing the reflectance. GRIN can be fabricated by gradually reducing the refractive index of the film from the refractive index of the substrate to the refractive index of air, as shown in Figure 3.8. However, achieving a unity refractive index at the film-air interface is practically impossible. The closest practical approach is reducing the packing density of the films which can negatively affect the mechanical strength of the ARCs and raise durability issues for solar panels [4].

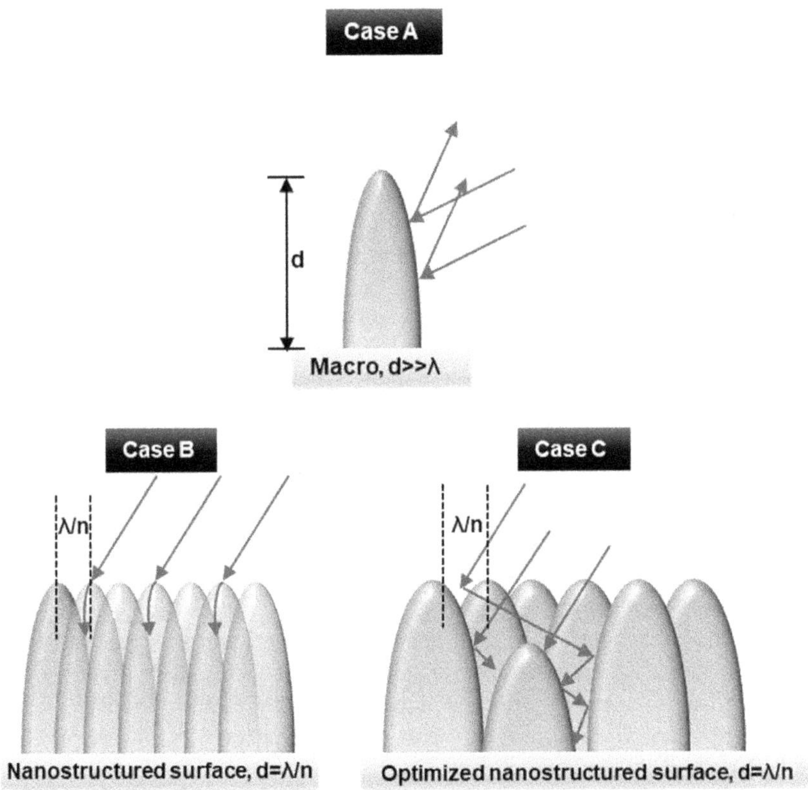

FIGURE 3.7 Light being reflected away from a macrostructure, light interacting with the whole rough surface due to comparable dimensions, and light undergoing multiple internal reflections through a nanostructure pattern.

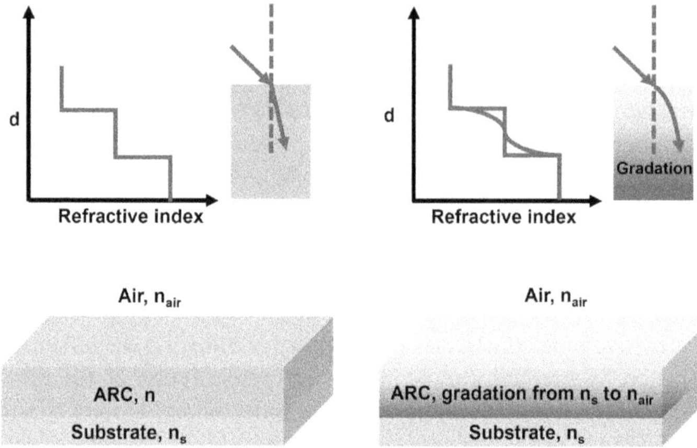

FIGURE 3.8 Schematic diagram representing single layer and gradient refractive layers.

3.5 OVERALL QUALITY, PERFORMANCE, EFFICIENCY, AND STABILITY OF SOLAR CELLS

The performance of solar cells can be reduced by the main three loss factors, namely optical loss, thermalization loss, and electrical loss, as shown in Figure 3.9. Optical losses stem from shading and surface reflections, with about 72% of incident light absorbed by the cell, diminishing short-circuit current density.

Mitigation strategies include applying anti-reflection coating, surface texturing, employing light trapping, and minimizing top contact coverage. Thermalization loss arises due to crystalline silicon's bandgap of approximately 1.1 eV. Tandem cells utilize multiple semiconductor materials with varying bandgaps stacked atop each other. This configuration allows the absorption of a broader spectrum of sunlight, reducing thermalization losses and enhancing overall efficiency. Recombination loss, stemming from inadequate passivation, diminishes the open circuit voltage significantly. To mitigate top surface recombination, a "passivating layer" can be applied. Compared to metals, silicon exhibits inferior current transport capabilities, resulting in elevated internal resistance and subsequent electrical losses. These losses negatively impact the fill factor (FF) of the cell. Implementing techniques to minimize these parasitic resistances is crucial for maximizing cell performance.

Thus, to achieve optimal efficiency, a solar cell necessitates an innovative structure, enhanced light absorption, abundant carriers with minimal recombination loss, reduced electrode resistance, and minimized electrode coverage area. Monocrystalline solar panels offer the highest efficiency, ranging from 15% to 22% in real-world conditions, but are expensive. In contrast, polycrystalline panels are more budget-friendly but offer slightly lower efficiency, typically between 12% and 17%. LONGI recently announced a new world record efficiency of 27.09% of silicon heterojunction back-contact (HBC) solar cells. Other advanced technologies such as Passivated emitter rear cell (PERC), Passivated emitter rear totally-diffused (PERT), Passivated emitter rear locally-diffused (PERL), Interdigitated back contact (IBC),

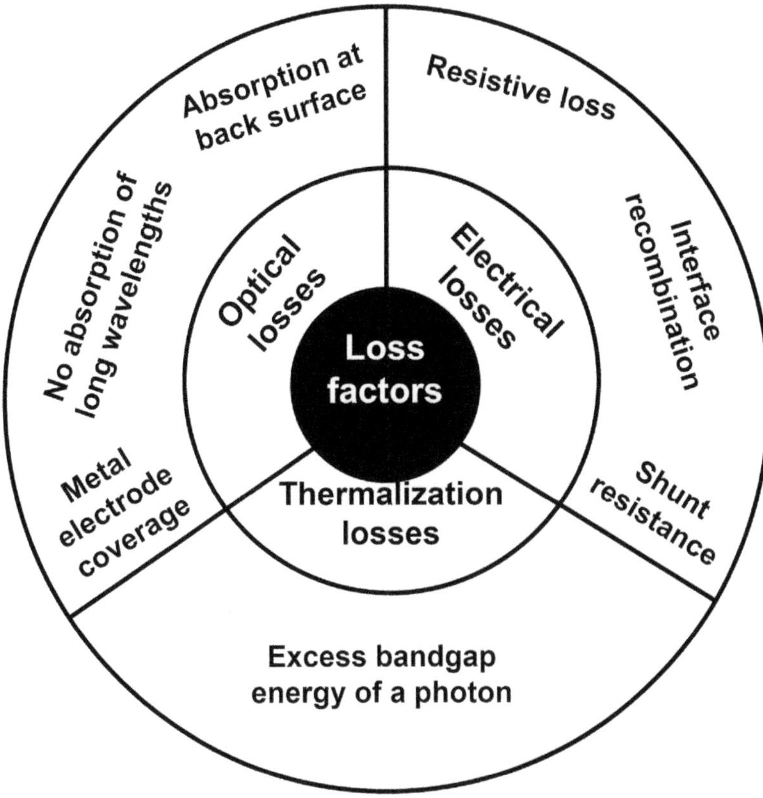

FIGURE 3.9 Loss factors in single junction solar cells.

and Heterojunction intrinsic thin-layer (HIT), have also emerged, notably enhancing efficiency and cost-effectiveness. Silicon solar cells have demonstrated excellent longevity and robustness due to the inherent stability of silicon. Most reputable solar panel manufacturers offer warranties ranging from 20 to 25 years, which evidences the stability and reliability of silicon solar panels [5].

3.6 DESIGN OF SOLAR PANEL

A silicon solar panel is designed by connecting individual solar cells together to increase their output power. Designing a silicon solar panel involves several key steps that can enhance the power output and lifetime performance to over 20 years of solar panels. Therefore, it is important to identify these factors while designing a silicon solar panel. The key steps are as follows:

Selection of silicon wafers: High-quality silicon wafers are chosen as the base material. These wafers are typically made from either monocrystalline or polycrystalline silicon.

ARC/texturing: A thin ARC may be applied to the front surface of the solar cell to minimize reflection losses and maximize light absorption. Silicon solar cells require a top surface that allows high transmission of light within the wavelength range of 350–1200 nm and minimizes reflection from the front surface. Although ARC theoretically offers a solution to reduce this reflection, often lacks the durability needed to withstand typical conditions in PV systems. An alternative approach involves "roughening" or texturing the surface, which can effectively decrease reflection and enhance light absorption. This process may entail treating the surface of silicon wafers to achieve the desired outcome.

Front contact formation: Thin metal contacts (usually made of silver) are applied to the front surface of the solar cell to collect the generated charges. The surface material atop a solar panel must be water-resistant, impact-resistant, UV-stable, and have low thermal resistance. Water ingress can corrode metal components, shortening the module's lifespan. Either the top or rear surface should be mechanically rigid to support cells and wiring, as the front surface typically provides strength.

Encapsulation: Encapsulation involves using a stable, optically transparent material like ethyl vinyl acetate (EVA) to adhere solar cells to the module's surfaces. EVA sheets are inserted between cells and surfaces, then heated to 150°C to polymerize and bond the module. Encapsulation with a transparent and durable material like glass or plastic protects against environmental damage.

Packing density: The packing density of solar cells within a PV module refers to the ratio of the module area covered by solar cells to the area left blank, as shown in Figure 3.10.

This density impacts both the module's power output and its operating temperature. The actual packing density is influenced by the shape of the solar cells utilized. For instance, monocrystalline solar cells typically have a round or semi-square shape, while polycrystalline silicon wafers are usually square. Consequently, if monocrystalline solar cells are not precisely cut to shape, the packing density of a monocrystalline module may be lower compared to that of a polycrystalline module. The comparison of packing densities achievable with round versus square cells is depicted below, where the round ingots-based wafers give a low packing density so the edges are cut off to produce semi-square cells and higher packing density.

Next, these cells are connected together (generally, in series) to increase the power and voltage compatible with a 12 V battery. The connection is made such

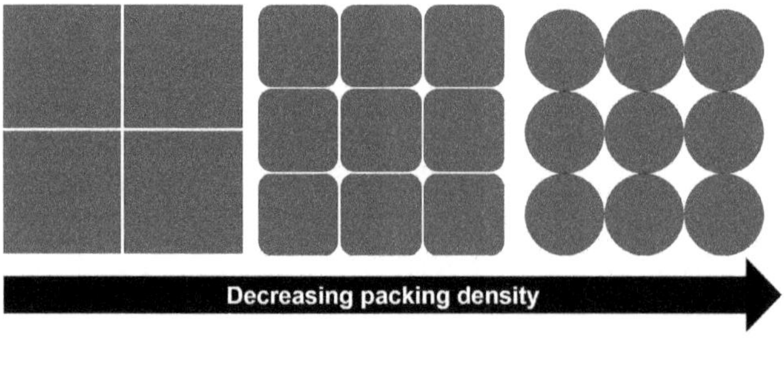

FIGURE 3.10 The different packing densities of silicon solar cells within a PV module.

that the mismatch effect can be avoided. The mismatch in PV modules occurs when the electrical parameters of one solar cell are significantly altered from those of the remaining devices leading to power loss. The factors on which this mismatch depends on are the operating point of the PV module, the circuit configuration, and differences in any part of the IV curve between one solar cell and another. One example of a mismatch is the short-circuit current density mismatch for cells connected in series, which can be caused by factors like shading/shadowing. Shading is a problem in PV modules since shading just one cell in the module can reduce the power output to zero. The short-circuit current for the combination is dragged down to the level of the poorer cell.

When the series string's current nears a faulty cell's short-circuit level, it restricts the overall current. Excess current forward biases good cells, but short-circuiting reverses this, leading to reverse bias across shaded cells and hot-spot heating. This overheating can cause damage like cell cracking or solder melting. A bypass diode can prevent this by connecting it in parallel with opposite polarity to a solar cell. During normal operation, it's reverse-biased, acting as an open circuit. Yet, if a cell becomes reverse biased, the diode conducts, diverting current from good cells and limiting the reverse bias across the faulty cell, preventing hot-spot heating.

Module assembly: Multiple solar cells are connected in series and/or parallel to form a solar module. These modules are then connected together to create a solar panel.

Throughout the process, quality control measures are implemented to ensure the efficiency and reliability of the solar panels. Additionally, ongoing research and development efforts continually refine the design and manufacturing processes to improve the performance and reduce the cost of silicon solar panels.

3.7 MULTIJUNCTION TANDEM SOLAR CELLS

To meet the demand for more solar energy, we must boost current PV tech and innovate for better, more efficient, and scalable options. Achieving multi-terawatt solar production hinges on the efforts of scientists and engineers to develop and implement new solar generation technology. Organic–inorganic halide perovskite solar cells (PSCs) have attracted widespread attention recently due to their high light absorption coefficient, long carrier diffusion length, and adjustable bandgap. These semiconductors with key optoelectronic properties have led to a new era of solar cells. Single-junction PSCs efficiency has surged from 3.8% to 26.1% in about 12 years [6]. Yet, in these cells, only a fraction of sunlight energy is harnessed, with many photons losing energy as heat, as shown in Figure 3.11.

Enhancing photon extraction efficiency from the solar spectrum is vital, involving reducing thermalization and carrier recombination. The thermalization loss occurs in silicon solar cells because crystalline silicon has approximately 1.1 eV of bandgap. Tandem solar cells (TSCs) are one approach to increase the number of bandgaps, which further increases the efficiency. PSCs stand out as top cell contenders in TSCs, thanks to their adjustable bandgap. PSCs can be stacked with medium- and narrow-bandgap cells like Si, narrow-bandgap PSCs, organics, Copper zinc tin sulfide (CZTS), and Copper indium gallium selenide (CIGS). Perovskite/c-Si tandems, in particular, hold near-term potential, leveraging solution

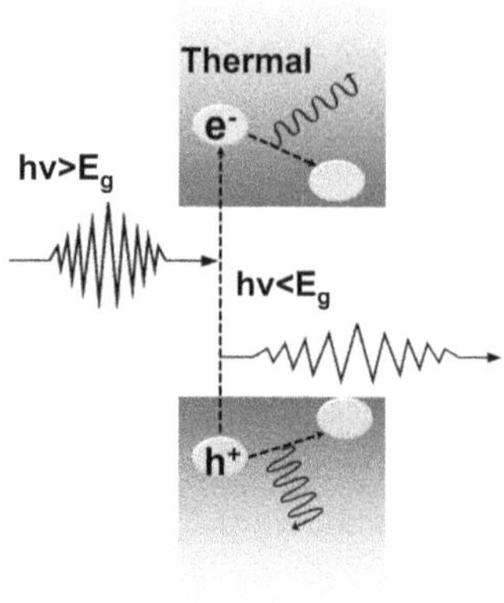

FIGURE 3.11 Schematic diagram of energy loss during photon absorption in a single-junction solar cell.

or vapor processing to outperform traditional Si cells for high performance. These tandems have rapidly evolved, with efficiencies climbing from 13.7% to 33.9% as recently announced by LONGI, though they still have room to grow towards the practical limit of 41%. Perovskite/silicon TSCs are typically configured in three ways: two-terminal (2T), where the cells are in direct series; three-terminal (3T), featuring an extra terminal for separate current extraction from each sub-cell; and four-terminal (4T), with electrically isolated cells, as shown in Figure 3.12.

The 4T setup enhances electrical independence between sub-cells, allowing for individual operation and optimization without requiring current matching, thus reducing constraints on top cell bandgap selection and making the device less sensitive to spectral changes [7].

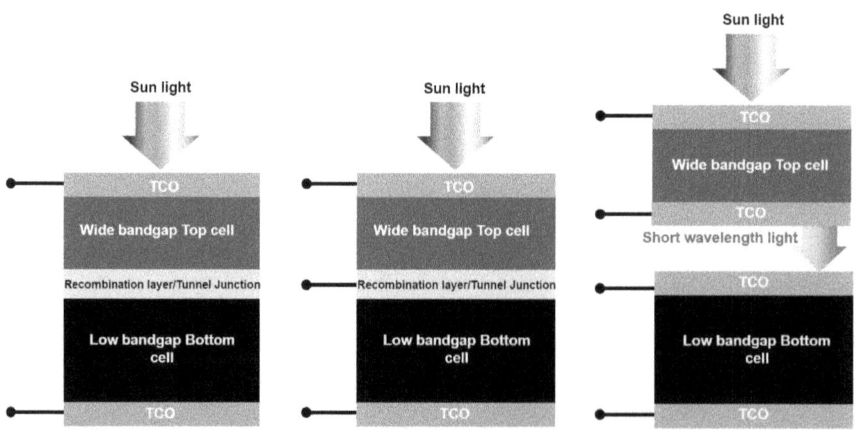

FIGURE 3.12 Schematic diagram of PSTSCs.

3.8 PROBLEM WITH SI SOLAR CELLS

Silicon wafer production presents intricate challenges, including maintaining purity and managing high costs. Solutions have emerged to streamline production effectively.

Purity maintenance: Ensuring silicon purity is critical for electronic device functionality. The Siemens and Czochralski processes create ultra-pure silicon. However, maintaining purity during slicing and polishing is challenging. Cleanroom environments and meticulous cleaning mitigate impurity introduction.

Cost reduction: High production costs stem from energy-intensive processes. Optimization enhances efficiency, such as controlling temperature and pulling rate in the Czochralski process. Silicon waste recycling minimizes material loss. Alternative materials like glass or plastic for non-critical applications also cut costs.

Space efficiency: Silicon solar cells have lower power density compared to some other types of solar cells, which means they require more space to generate the same amount of electricity. This can be a limitation in applications where space is constrained.

Efficiency: Silicon solar cells have a theoretical efficiency limit, and reaching this limit in practical applications remains a challenge. Improving efficiency involves addressing issues such as recombination losses, light trapping, and optimizing material properties.

Light absorption: A typical solar panel structure consists of glass for wind protection, a PV cell with an ARC in the middle, and a supportive metal layer at the bottom. Each layer's optical properties are described by a complex refractive index ($N=n-ik$), where n is the real refractive index and k is the extinction coefficient. The absorption factor (α) of a solar cell is determined by $\alpha=4\pi k/\lambda$, indicating that as photon wavelength increases, absorption decreases. Short-wavelength photons ($\lambda < 1.0\,\mu m$) possess sufficient energy to create electron-hole pairs due to their energy exceeding the band gap, making them readily absorbed by crystalline silicon, typically found in the infrared spectrum. Conversely, long-wavelength photons ($\lambda > 1.2\,\mu m$) lack the energy to generate electron-hole pairs and are poorly absorbed, rendering them underutilized by the solar cell. Photons in the transitional range ($1.0\,\mu m < \lambda < 1.2\,\mu m$) are weakly absorbed.

Despite these challenges, cost-cutting measures sustain silicon wafer production's economic viability amidst industry demand. Manufacturing complexity varies among solar panels, typically converting 15%–22% of solar energy into usable energy. During blackouts or overnight, a house can rely on up to 22% of available energy. Primary types include thin-film, monocrystalline silicon, and polycrystalline silicon panels. Monocrystalline silicon panels, crafted from high-grade silicon, are the most efficient. Balancing efficiency and cost pose a challenge, motivating engineers and scientists to enhance cell efficiency and reduce production costs for more economically viable solar energy. Panel prices hinge on various factors, including other component costs.

3.9 MARKET STATUS AT THE NATIONAL AND INTERNATIONAL LEVELS

The growing use of renewable energy sources because of their advantages for the environment and rising power demand is predicted to drive market expansion. Some technologies have vastly exceeded the expected trends, e.g., the passivated emitter and rear cell and the use of gallium as the dominant p-type dopant. In 2012, multi-crystalline silicon wafers represented over 60% of the solar cell market, which significantly increased between 2012 and 2015. The dominance of multi-crystalline wafers during that period was related to the lower processing costs associated with directional solidification, lower susceptibility to Boron-oxygen related light-induced degradation (BO-LID), and higher packing factor of square wafers in solar modules [8].

The market shares for directionally solidified silicon wafers reduced from 2018, after which the trends rapidly changed in favor of monocrystalline silicon. The market share of directionally solidified silicon wafers was approximately 3% in 2022, despite predictions of 2022 market shares of 10%–45%. This highlights that the industry shifted toward monocrystalline silicon much faster and to a broader monocrystalline silicon usage than predicted. However, the transition from PERC to Tunnel oxide passivating contact (TOPCon) might be even faster, like the transition from Al-BSF to PERC. This rapid transition is illustrated in recent announcements by leading manufacturers. JinkoSolar shipped 17.8 GW of solar modules in the second quarter of 2023, of which 10.4 GW (58%) were n-type modules. Although total module shipments increased by 36.2% compared with the previous quarter, n-type module shipments increased by 74.1%. Trina Solar cell capacity was forecasted to reach 75 GW by the end of 2023, with 40 GW based on n-type TOPCon cells. However, there have been significant discrepancies between the historically projected market shares compared to the current market trends for different silicon solar cell architectures. The global solar cell market is expected to reach a compound annual growth rate (CAGR) of 17.2% over the forecast period from 2023 to 2032. The solar cell market size was estimated at USD 108.8 billion in 2022 and it is projected to surpass around USD 508.18 billion by 2032. The growth of this sector also depends on a number of policies and incentives provided by various governments to boost the usage of solar systems, such as promoting industrial expansion, including tax breaks, subsidies, net metering, financial aid, reduced import duties, and feed-in tariffs. The market is anticipated to be driven by rising demand for these cells in the commercial and utility sectors. Over the projection period, declining PV prices are anticipated to expand the market for crystalline silicon modules. Furthermore, the demand for thin film solar cells in residential and commercial applications is anticipated to rise as a result of technical developments aimed at enhancing operational efficiency [8,9].

Chinese PV companies are aggressively entering growing markets by acquiring international manufacturers and constructing factories abroad, favorably affecting the size of the solar cell industry. By 2023, the Indian government wanted to generate 40 GW of electricity using solar technology. In 2023, manufacturing capacity expansions were mainly due to the expected reinstatement of the Approved List of Models and Manufacturers (ALMM) order in April 2024 and potential export opportunities. By December 2023, the country's cumulative solar module manufacturing capacity reached 64.5 GW, with solar cell manufacturing capacity totaling 5.8 GW [10,11]. The top ten manufacturers held 62% of module production capacity and the entirety of cell production capacity. Approximately 60% of module manufacturing capacity could produce M10 and G12 wafer sizes, with only 22.2 GW enlisted under the ALMM order. Monocrystalline modules comprised 67.5% of module production capacity, followed by polycrystalline, TOPCon, and thin film modules. Projections indicate module capacity exceeding 150 GW and cell capacity over 75 GW by 2026. Monocrystalline modules are expected to dominate both module (59.7%) and cell (50.5%) production capacities by 2026, with TOPCon, Heterojunction (HJT), and other technologies following suit [8–10]. Indian manufacturers expanding their solar panel production capacities must navigate geopolitical tensions and trade disputes, as cheaper Chinese products pose challenges to local module competitiveness.

3.10 CONCLUSIONS

This chapter highlights the essential role of silicon wafers in various aspects of human life and technological advancement, prized for their stability among semiconductor materials. Not only are these wafers a superior alternative to metallic substances, but they are also abundant on Earth. Research on silicon has propelled significant technological progress, notably with the invention of ICs, which have transformed manufacturing and turned large machinery into portable devices. Ongoing research aims to expand wafer size and control properties through doping, promising more sophisticated innovations from silicon wafers in the near future. Recognized for their unique electrical and thermal properties and high purity, these wafers are indispensable for diverse applications. As a fundamental component in the technology industry, silicon wafers have also revolutionized the PV sector. Silicon solar cells have the potential to be combined in tandem configurations with other PV technologies like PSCs, offering solutions to today's energy challenges. The national and international market potential of the silicon-based PV industry has grown in recent decades and is expected to further expand, contingent upon ongoing technological advancements and supportive government policies.

NOTE

* Corresponding author: Barkha Tyagi, Department of Physics, University of Oxford, Oxford, United Kingdom.

REFERENCES

1. A. Ayodele, *Silicon Wafers: Production, Properties and Application*, Wevolver, 2024. https://www.wevolver.com/article/silicon-wafers.
2. L. Bourget, G. Brucker, M. Feaver, G. Hill, Y. Ichihashi, K. Koai, A. Larson, C. L. Tiec, L. Martinez, I. Pokidov, A. Radomski, G. Rosenzweig, E. Rotem, C.-Y. Tai, M. Tricard, M. V. Spyk, and T. K. Whidden, *MKS Instruments Handbook: Semiconductor Devices & Process Technology*, 2nd edition, 2023. MKS, Andover, Massachusetts.
3. H. Wang, C. Zhang, W. Chan, S. Tiwari, F. Rana, Ultrafast response of monolayer molybdenum disulfide photodetectors, Nature Communications, 6, 8831 (2015).
4. H. K. Raut, V. A. Ganesh, A. S. Nair, and S. Ramakrishna, Anti-reflective coatings: A critical, in-depth review. *Energy & Environmental Science* 2011, 4(10), 3779–3804.
5. S. Chowdhury, M. Kumar, S. Dutta, J. Park, and J. Kim, High-efficiency crystalline silicon solar cells: A review. *New Renewable Energy* 2019, 15, 3.
6. NERL Transforming Energy Group, *Best Research-Cell Efficiency Chart*, 2024. https://www.nrel.gov/pv/cell-efficiency.html.

7. Y. Shi, J. J. Berry, and F. Zhang, Perovskite/silicon tandem solar cells: Insights and outlooks. *ACS Energy Letters* 2024, 9, 1305–1330.
8. B. V. Stefani, M. Kim, Y. Zhang, B. Hallam, M. A. Green, R. S. Bonilla, C. Fell, G. J. Wilson, and M. Wright, Historical market projections and the future of silicon solar cells. *Joule* 2023, 7(12), 2684–2699.
9. B. J. Albertus, K. Dummit, D. Feldman, S. Grossman, and J. Zuboy, *The State of the Solar Industry*, US Department of Energy, Solar Energy Technology Office, 2024. www.energy.gov/sites/default/files/2024-04/Wed%20AM%20-%20Solar%20industry%20update.pdf
10. Power Engineering International. *Report Shows Bright Future for India's Solar PV Manufacturing Market*, Power Engineering International, 2024.
11. K. Batool, Z.-Y. Zhao, M. Irfan, S. Ullah, and C. Işik. Assessing the competitiveness of Indian solar power industry using the extended Five Forces Model: A green innovation perspective. *Environmental Science and Pollution Research* 2023, 30, 34, 82045–82067.

4 Fabrication Techniques for Perovskite and Organic Solar Cells

*Neetesh Kumar**

4.1 INTRODUCTION

The most abundant source of energy in the universe is solar energy. Devices that convert solar radiation into usable electrical energy are known as solar cells. This term typically refers to individual units. When solar cells are combined in series and parallel configurations, they form solar panels. Photovoltaic (PV) systems, which generate electricity without carbon dioxide emissions, are becoming increasingly popular among homeowners for rooftop installations, either to replace or supplement conventional electric supply. Solar panels are also widely used in remote terrestrial locations where traditional power sources are unavailable or costly to install. Unlike wind turbines and hydroelectric generators, solar panels have no moving parts, reducing the need for maintenance. This makes them ideal for use in space installations, such as satellites and space stations. In addition to traditional silicon-based solar cells, modern photovoltaics technology has seen significant advancements, particularly with the development of perovskite solar cells and organic solar cells. These new technologies promise to revolutionize the solar energy industry by offering higher efficiencies, lower production costs, and increased versatility.

Perovskite solar cells (PSCs) are a type of photovoltaic device that converts sunlight into electricity through the photovoltaic effect. They are composed of thin layers of perovskite-structured materials, typically metal halides like lead iodide, methylammonium lead halide, or mixed perovskites. These materials exhibit excellent light-absorbing properties, enabling efficient conversion of solar energy into electrical power. Within a short period of intensive research, the power conversion efficiency (PCE) of PSCs has remarkably surpassed 26% [1–3]. Notably, achieving such high efficiency through solution processing is unprecedented in other PV technologies, positioning PSCs as promising candidates for high-performance, cost-effective PV systems. On the other hand, organic polymer-based photovoltaic devices, also known as organic solar cells (OSCs), offer a more promising alternative to traditional silicon-based solar cells, with their flexibility, lightweight design, and potentially lower manufacturing costs. Recent years have witnessed significant strides towards large-scale manufacturing of PSCs and OSCs using scalable deposition methods. Notably, there has been notable progress in enhancing the efficiency across different size scales, including small-area cells (~0.1 cm^2), large-area cells (~1 cm^2), and modules (>10 cm^2) [4]. The advancements in small-area devices

DOI: 10.1201/9781003499701-4

primarily stem from refined active layer composition and morphology, driven by innovations in solution chemistry and fabrication techniques. These breakthroughs have seamlessly transitioned to larger devices, resulting in efficiency gains that mirror the progress observed in smaller-scale developments. Scaling up these PSC and OSC devices presents several challenges, such as devising deposition methods capable of uniformly coating all device layers across large-area substrates. This includes the photoactive layer, electron and hole transport layers (ETL and HTL), and electrodes. Additionally, enhancing control over film formation across the entire device stack on a larger scale involves refining precursor chemistry to align with processing techniques. There's a need to establish protocols for producing and integrating modules while investigating how device architecture influences module interconnections, reliability, and operational stability.

The strides made in scaling up PSCs and OSCs are notable; small-area laboratory devices created through scalable deposition methods like blade coating [5,6] and slot-dye coating [7,8] have nearly matched the PCE levels of those produced via spin coating. While module efficiency still trails that of smaller devices, the rate of improvement in module PCE is unprecedented in comparison to other PV technologies. Recently Nazeeruddin et al. [9] demonstrated a perovskite module, certified with a 16% PCE and an area exceeding 27.22 cm^2, whereas Basu et al. [10] have demonstrated an OSC module with 14.5% PCE on 204 cm^2 active area. These achievements signify significant progress toward the commercial viability of large-scale PSCs and OSCs.

This chapter provides an overview of common fabrication techniques including solution and non-solution-based techniques. Furthermore, it offers a comparative analysis of various scalable deposition methods from solution-phase to vapor-phase depositions-highlighting their respective advantages, limitations in PSC processing, and the current status of efforts to scale up PSCs and OSCs. These techniques are designed to leverage the distinctive properties of these materials, all with the overarching goal of achieving efficiency, cost-effectiveness, and scalability in their production.

4.2 SOLUTION PROCESSING-BASED TECHNIQUES

Numerous techniques have been developed for depositing the perovskite and polymer films for the fabrication of PSCs and OSCs. Solution processing-based techniques have emerged as powerful methods for the fabrication of these devices offering advantages in scalability, cost-effectiveness, and flexibility. These techniques utilize solution-based approaches to deposit active materials onto substrates, enabling the production of thin-film solar cells with high efficiency and low manufacturing costs.

4.2.1 Spin-Coating Technique

Spin coating is a widely used solution-based approach and is utilized for the uniform deposition of perovskite and polymer layers on the rigid and flexible substrates. This process can be broadly divided into four main steps: deposition, spin up, spin-off, and evaporation. After the coating process, a subsequent annealing step is employed to boost the crystallization of the perovskite film and promote strong bonding between metal cations and halogen anions. Fine-tuning the film's thickness and quality is achieved by adjusting parameters such as spin speed, acceleration, and duration. Notably, the highest laboratory-scale efficiency achieved through this method stands at approximately 26.07% [1,3]. This approach is applicable to both inverted and regular PSC [OSC structures] [11,12]. Both one-step and two-step-based perovskite films can be fabricated by this technique. Despite its notably high efficiency, spin-coating's limitation lies in creating uniform films over larger surfaces. Spin coating boasts several key advantages and disadvantages:

Ease of setup and uniform coating: It offers a straightforward setup process and allows for thin and consistent coatings at various thicknesses, making it suitable for research and rapid prototyping purposes.

Rapid drying and consistency: The ability to use high spin speeds results in quick drying times due to increased airflow. This rapid drying enhances consistency at both macroscopic and nano length scales, often eliminating the need for additional heat treatment after deposition.

Cost-efficiency for individual substrates: It is a cost-effective method for batch processing individual substrates. In contrast to other techniques requiring pricier equipment and higher energy inputs, spin coating remains relatively low-cost.

One primary disadvantage of spin coating is its limitation in scaling up for large-scale production, the film thickness to differ from the center to the edges of the substrate. While effective for individual substrates and small-scale applications, the process may become impractical when aiming for mass production due to its batch-oriented nature and the challenges associated with achieving uniform coatings across larger areas. The one and two-step perovskite deposition process using the spin-coating technique is illustrated in Figure 4.1.

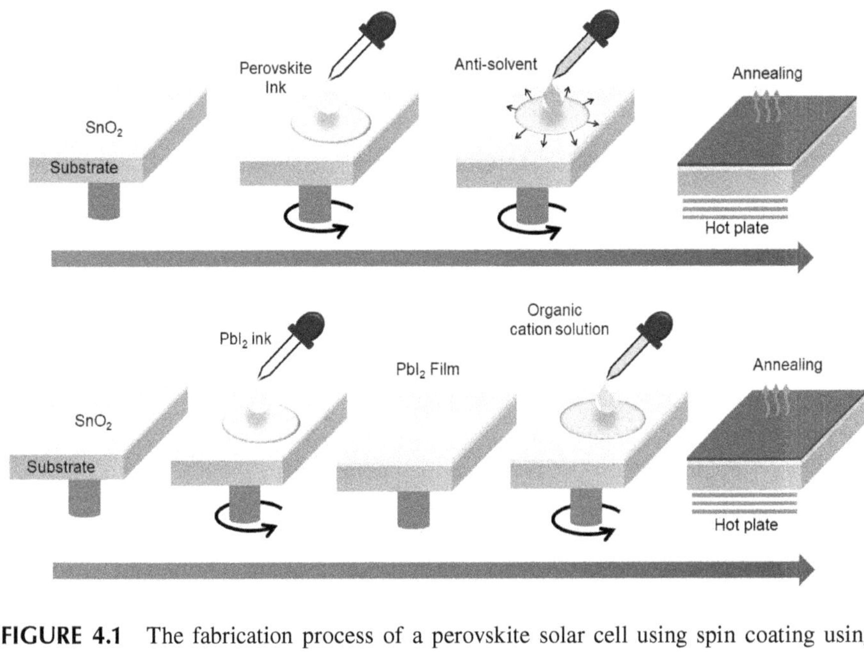

FIGURE 4.1 The fabrication process of a perovskite solar cell using spin coating using one-step (top) and two-step process (bottom).

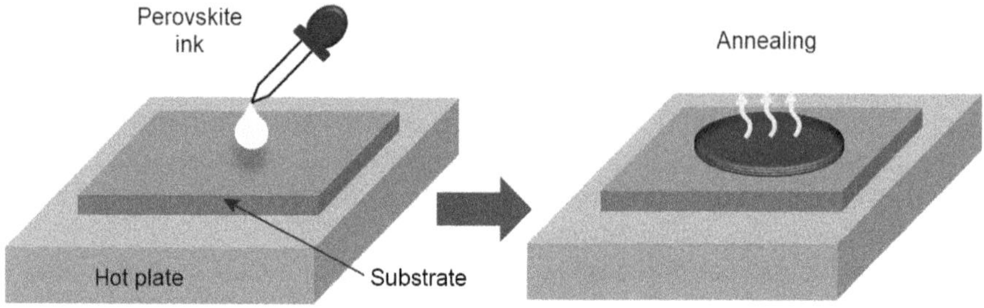

FIGURE 4.2 The fabrication process of a perovskite layer using drop-casting.

4.2.2 Drop-Casting

Indeed, drop casting stands as a foundational and cost-efficient method in the manufacturing of both PSCs and OSCs. Unlike spin coating, it notably diverges by eliminating the need for substrate spinning. Instead, the film's thickness and properties rely on the volume and concentration of the perovskite or polymer precursors in the solution. Other aspects like substrate wettability, evaporation rate, and the drying process also influence the film's quality. Typically, volatile solvents that effectively wet the substrate are preferred for this technique. Compared to spin coating, drop casting reduces material wastage, yet it presents challenges, such as difficulties in controlling film thickness and the creation of uneven films on larger substrates. Ding et al. used a drop-casting technique to make $MAPbI_3$ and $CsPbI_{2.25}Br_{0.75}$ films for solar cells with an efficiency of over 18.0% [13,14]. Recently, the preparation of hybrid perovskite films comprising both quasi-2D and quasi-3D phases by this technique has been intensively researched due to the unique type of crystal growth as illustrated in Figure 4.2 [15,16].

4.3 ROLL-TO-ROLL PRINTING TECHNIQUES

Roll-to-roll (R2R) printing techniques have gained significant attention in the fabrication of solar cells due to their potential for high-throughput, low-cost manufacturing. These techniques offer advantages such as scalability, versatility in material choice, and the ability to produce large-area devices. This technique is extensively used for PSC and OSC manufacturing. Within this method, diverse coating approaches assist in creating thin films on moving substrates using rotating rolls, rods, or slot-dyes etc. R2R printing techniques have indeed extended the possibilities of manufacturing processes by leaps and bounds, often providing efficiencies and cost reductions by up to 200% or more compared to traditional batch processing methods. This dramatic increase in efficiency stems from several key factors:

High throughput: R2R printing allows for continuous, high-speed production of flexible materials, enabling manufacturers to produce large quantities of products in a relatively short amount of time. Compared to batch processing, which requires frequent stopping and starting, R2R techniques can achieve significantly higher throughput rates.

Reduced material waste: R2R printing minimizes material waste by optimizing ink, substrate, and energy usage. The continuous nature of R2R processes ensures that materials are utilized more efficiently, resulting in less waste compared to batch processes where setup and cleanup phases can lead to significant material losses.

Lower labor costs: R2R printing requires less manual intervention and supervision compared to batch processing methods. Once the printing parameters are set, the process can run autonomously for extended periods, reducing the need for labor-intensive tasks and lowering labor costs.

R2R printing techniques are highly scalable, making them suitable for both small-scale Scalability and large-scale production. Manufacturers can easily adjust production volumes to meet demand without significant changes to the printing setup, allowing for greater flexibility and agility in responding to market fluctuations.

Cost savings: By increasing throughput, reducing material waste, and optimizing labor costs, R2R printing techniques can lead to significant cost savings for manufacturers. These savings can be as high as 200% or more compared to traditional batch processing methods, making R2R printing an attractive option for industries looking to improve efficiency and competitiveness.

Overall, the adoption of R2R printing techniques has revolutionized manufacturing processes across various industries, enabling faster, more cost-effective production of a wide range of products, from flexible electronics and solar cells to flexible packaging and wearable devices. As technology continues to advance, R2R printing is expected to play an increasingly vital role in shaping the future of manufacturing.

4.3.1 Spray Coating

The spray coating technique employs a nozzle to dispense tiny liquid droplets onto the substrate. The size and placement of these droplets may vary, requiring several droplets to overlap for complete substrate coverage. This technique can be categorized into three types based on the method used to control spray droplets: pneumatic spraying (utilizing pressurized gas flow), ultrasonic spraying (utilizing ultrasonic vibrations), and electro-spraying (utilizing electrical repulsion). To achieve well-controlled and uniform film coating, it is crucial to regulate the solvent evaporation rate during the spray coating process. Figure 4.3 presents a schematic image illustrating the pneumatic spray-coating process.

In this approach, compressed gas (such as N_2 or Ar) exerts significant pressure on the coating ink, propelling it out of the nozzle where it swiftly interacts with the surrounding gas. This interaction leads to both deceleration and dispersion of the coating ink due to air resistance, resulting in the formation of a fine mist before it reaches the substrate. However, the robust gas flow in pneumatic spray systems may potentially displace active coating ingredients in the liquid phase, causing notable material losses and reducing its suitability for large-scale production. Nevertheless, this method remains versatile and adaptable. Some systems utilize specialized nozzle shapes to convert the coating fluid into a fine mist, facilitating uniform coating on uneven surfaces. Others are engineered to maintain a high spray rate, ensuring consistent and even coating application.

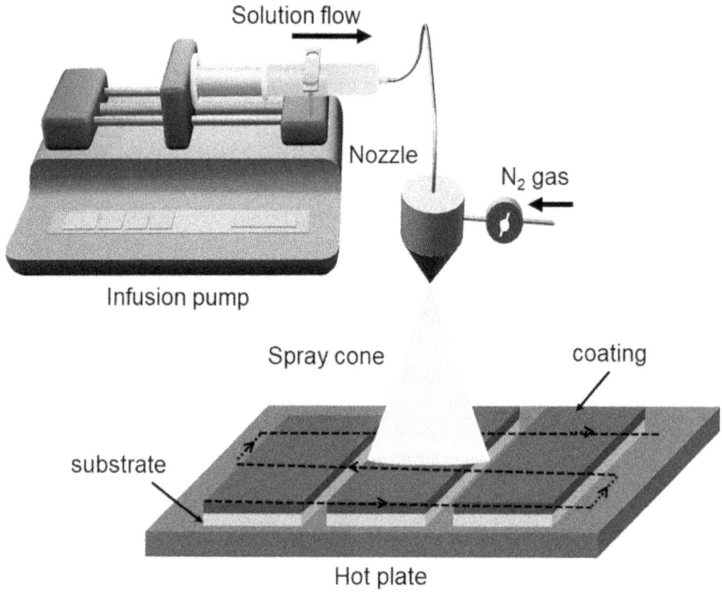

FIGURE 4.3 The fabrication process of a perovskite layer using pneumatic spray coating on a large area with a movable spray head.

Spray-coating methods can be categorized into three distinct modes of operation: (a) stationary spray, where the spray head remains fixed; (b) single-pass spray, where the spray head moves in only one direction; and (c) raster spray, where the spray head is movable along both the x and y axes. The majority of reported spray-coating techniques for perovskite deposition utilize either stationary spray or single-pass spray modes. Film thickness can be tuned by increasing the spray time or spray pass. This technique is adaptable for depositing both planar and inverted solar cells.

Using a simple airbrush for the spray deposition of perovskite layers has proven to be viable for PSC fabrication [17,18]. Additionally, many groups have also reported the combined use of spin coating (first step) and airbrush spraying (second step) in a two-step fabrication sequence for perovskite thin films [18–20].

While spray coating is recognized as a scalable deposition method, the majority of documented studies focus on small cells with active areas typically equal to or less than 1 cm^2. The efficiency of spray-deposited PSC depends on several parameters, such as device structure and perovskite absorber composition. However, precise control over processing parameters such as carrier gas, substrate temperature, and annealing temperature remains crucial for device performance. The recorded efficiency using one-step [21] and two-step spray-coated [22] perovskite thin films is approximately 18.3% and 20.5%, respectively. Spray deposition is also used for the deposition of other layers of PSCs [23,24], and OSCs [25].

There are several challenges associated with the spray deposition technique:

Uniformity: Achieving uniform deposition across the substrate can be challenging, especially for large areas. Variations in spray parameters, such as spray pressure and nozzle-to-substrate distance, can result in non-uniform film thickness and coverage.

Edge effects: Spray deposition often leads to thicker coatings near the edges of the substrate due to overspray and uneven distribution of droplets. This can result in edge defects and non-uniform film properties.

Substrate compatibility: The spray deposition technique may not be suitable for all types of substrates. Certain substrates may be prone to damage or deformation due to the high velocity of the spray droplets or the solvents used in the spray solution.

Material waste: Spray deposition can result in material waste, particularly during the setup and optimization stages where multiple iterations may be required to achieve the desired film quality. Additionally, overspray can lead to inefficient use of materials.

Process control: Controlling the spray parameters, such as spray velocity, droplet size, and distribution, is crucial for achieving desired film properties. However, maintaining precise control over these parameters can be challenging, especially in large-scale production environments.

Drying and solvent evaporation: Proper drying and solvent evaporation are critical for achieving high-quality films. However, achieving uniform drying across large areas can be difficult, leading to variations in film morphology and properties.

Surface tension effects: Surface tension effects can cause non-uniform spreading of the spray solution on the substrate surface, leading to uneven film coverage and defects.

Addressing these challenges requires careful optimization of spray parameters, substrate preparation, and process conditions. Additionally, alternative deposition techniques or process modifications may be necessary to overcome specific limitations associated with spray deposition.

To tackle these issues, modified spraying processes like the spray-coating technique [26], electrostatic pulsed spray coating [27], airbrush pen spray [28], and ultrasonic spray-coating technique [29] are employed.

4.3.1.1 Ultrasonic Spray Coating

Ultrasonic spray technology indeed employs ultrasonic energy to break a liquid into very fine droplets, typically ranging from a few microns to tens of microns in size. This technology finds applications in various industries including coatings, pharmaceuticals, electronics, and biotechnology. Ultrasonic spray systems are equipped with a chip, known as the atomization surface, positioned at the nozzle tip. The mechanical vibrations generated by ultrasonic waves cause the coating fluid to surround the chip, inducing agitation on the surface. As the ultrasonic output exceeds the surface tension, the fluid is emitted from the surface in the form of a fine mist, resulting in the production of ultrafine and uniformly sized droplets. This capability is particularly advantageous for coating small areas consistently. The size of the generated droplets, typically in the micrometer range, is dictated by the specific resonant frequency (usually between 20 and 200 kHz) at which the ultrasonic nozzle operates, with higher frequencies yielding smaller median droplet sizes. Furthermore, the system minimizes material waste by reducing splashing and unnecessary dispersion of the coating fluid. This efficiency is achieved as the gas flow in an ultrasonic spray solely directs the mist of droplets without atomizing them further. Spray nozzles come in a variety of lengths, shapes, and sizes to accommodate diverse applications. Precise control over the thickness, uniformity, and quality of the film coating is attainable by adjusting the flow

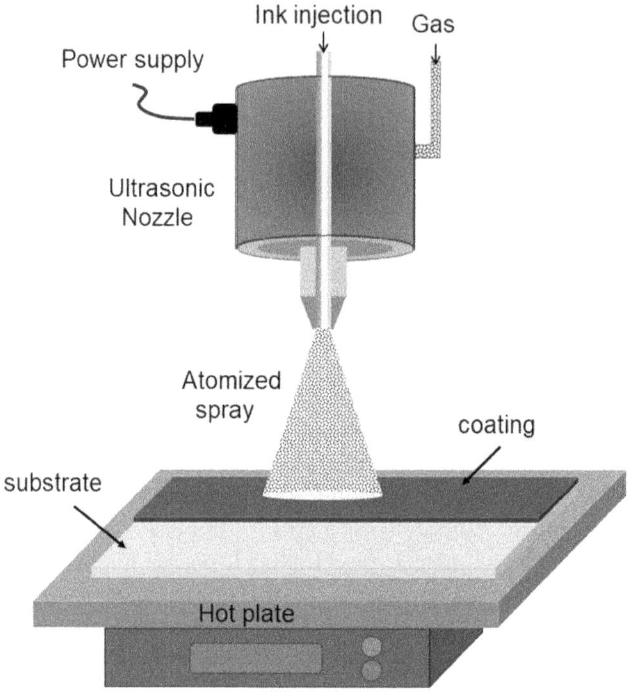

FIGURE 4.4 The fabrication process of a perovskite layer using ultrasonic spray coating.

rate, vibrational frequency, and dispensing pathway through an automated system. Overall, ultrasonic spray technology offers advantages such as precise control over droplet size, uniform coating deposition, reduced overspray, and efficient use of materials. These characteristics make it a preferred choice for various applications requiring precise and uniform spray coatings. The fundamental schematic for the ultrasonic spray-coating technique is depicted in Figure 4.4.

The conventional spray-coating technique typically results in approximately 10%–12% lower PCE compared to the ultrasonic spray-coating technique. This discrepancy arises from the ultrasonic method's ability to precisely regulate the deposition rate and enhance film quality. Films produced through ultrasonic spray-coating tend to have larger grain sizes, which improve carrier mobility and reduce charge recombination, ultimately leading to higher efficiency. Additionally, these films exhibit reduced hysteresis in comparison to those produced through spin-coating and conventional spray-coating methods. Consequently, the ultrasonic spray-coating technique emerges as the preferred method for uniformly depositing the perovskite layer [30–32] passivating layer [33], 2D perovskite [34]. The highest efficiency of an area PSC device fabricated with ultrasonic spray coating is 20%, whereas a PSC module of 112 cm^2 active area achieved a PCE of 13.8% [30].

4.3.1.2 Blow Drying

The blow-drying methodology plays a pivotal role in the intricate process of fabricating perovskite solar cells, a cutting-edge technology with immense potential in renewable energy applications. At the heart of this methodology lies a meticulous sequence of steps aimed at producing high-quality perovskite films crucial for achieving efficient photovoltaic conversion. To commence, the process initiates with the application of a perovskite precursor solution onto the designated substrate. This solution, laden with the essential components for perovskite formation, is delicately dispensed onto the substrate surface utilizing advanced deposition techniques such as spin coating, blade coating [35], bar coating [36], slot die coating [37], etc. This initial step sets the foundation for subsequent stages, laying the groundwork for the formation of the perovskite film. Following the precise application of the precursor solution, the blow-drying methodology comes into play, constituting a critical phase in the fabrication process. A controlled stream of gas, typically nitrogen or air, is strategically directed towards the substrate. This stream serves a dual purpose: to hasten the evaporation of the solvent present within the precursor solution and to promote the formation of a uniform perovskite film. As the gas stream gently envelops the substrate, it facilitates the evaporation process by swiftly removing the solvent molecules from the precursor solution. This rapid evaporation is essential for preventing the formation of undesirable defects within the perovskite film, ensuring its integrity and structural uniformity. Additionally, the directed flow of gas aids in the redistribution of perovskite components, promoting the formation of a homogeneous film with a consistent thickness across the substrate surface. The culmination of the blow-drying process yields a thin film composed of perovskite material firmly adhered to the substrate. This film, characterized by its uniformity and absence of defects, serves as the active layer within the perovskite solar cell device. Its impeccable structural integrity is paramount for facilitating efficient charge transport and light absorption, thereby maximizing the device's photovoltaic performance.

In essence, the blow-drying methodology stands as a linchpin in the fabrication of perovskite solar cells, bridging the gap between precursor solution deposition and the realization of high-performance photovoltaic devices. Through meticulous control and optimization of this methodology, researchers and engineers strive to unlock the full potential of perovskite-based technologies, driving forward the advancement of sustainable energy solutions. Figure 4.5 shows the simplest blow-drying fabrication of perovskite films.

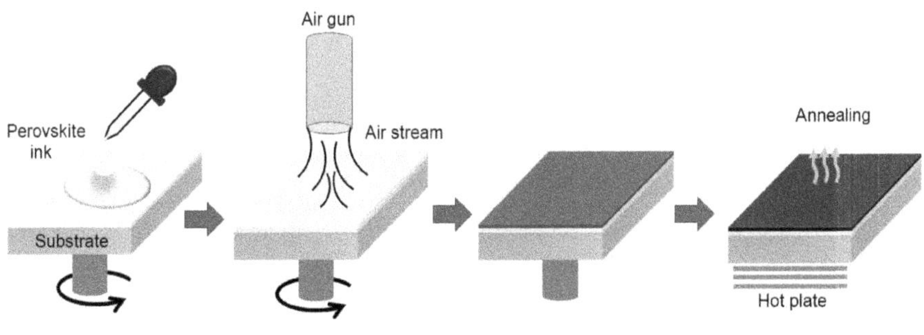

FIGURE 4.5 Blow drying technique for perovskite film fabrication.

4.3.2 Blade Coating

Blade coating of perovskite film is a sophisticated technique employed in the fabrication of perovskite solar cells, offering precise control over the deposition process to achieve uniform and high-quality thin films. Figure 4.6 illustrates the blade coating process used for the fabrication of perovskite films. This method involves the application of a perovskite precursor solution onto a substrate using a sharp blade or a doctor blade, which spreads the solution evenly across the substrate surface while simultaneously controlling the film thickness. The blade glides smoothly across the substrate surface, it delicately spreads the precursor solution, ensuring uniform coverage and thickness control. The gap between the blade and the substrate is finely calibrated, dictating the precise thickness of the resultant perovskite film. This exquisite level of control is instrumental in the fabrication of perovskite films with unparalleled uniformity, free from imperfections and defects that could impede device performance.

Simultaneously, as the precursor solution is evenly distributed across the substrate, the delicate dance of solvent evaporation and perovskite crystallization begins. This transformative phase marks the transition from a wet film to a solid-state thin film, characterized by the emergence of crystalline structures crucial for efficient photovoltaic conversion. The blade coating technique, with its unparalleled precision and finesse, facilitates the nucleation and growth of perovskite crystals in a controlled manner, ensuring the formation of high-quality films optimized for photovoltaic applications. Throughout the drying and crystallization process, meticulous attention is paid to environmental conditions, with factors such as temperature, humidity, and airflow carefully regulated to foster optimal film formation. This holistic approach ensures the creation of perovskite films endowed with the desired structural and optoelectronic properties, laying the foundation for high-performance solar cell devices.

Upon completion of the blade coating process, the substrate adorned with the pristine perovskite film may undergo further processing steps, including the deposition of additional functional layers such as electron and hole transport materials.

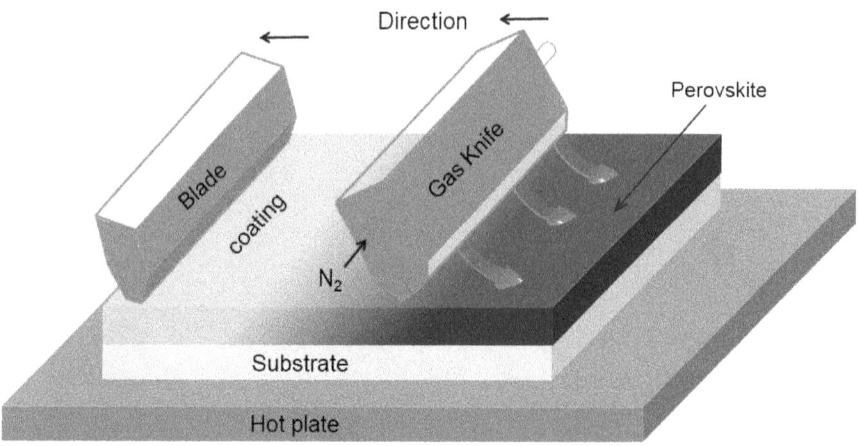

FIGURE 4.6 Blade coating of perovskite ink and blow drying to convert into perovskite.

These ancillary layers serve to enhance charge transport and collection within the device, further optimizing its photovoltaic performance.

Compared to spray-coating and other roll-to-roll compatible methods, this technique ensures enhanced control over film morphology. It is applied in crafting planar heterojunction PSCs. Generally, most perovskite deposition approaches result in pinhole formation due to uneven deposition, moisture, and air exposure, leading to diminished performance of photovoltaic devices. To combat these issues, blade coating is the preferred choice. It is highly recommended for fostering self-assembled perovskite crystalline domains, ensuring uniform film coverage and improved device stability. Additionally, this technique offers the benefits of uniform, high-speed deposition and easy application on flexible substrates [38]. This technique is commonly used as a single-step perovskite film fabrication on small and large-scale [39–42]. Two-step blade-coated PSCs also achieved PCE of over 20% and module of 13% ($10 \times 10\,cm^2$) [43].

4.3.3 Slot Die Coating

Slot die coating of perovskite film stands as a pinnacle of precision and innovation in the intricate landscape of perovskite solar cell fabrication. This highly sophisticated method offers unparalleled control over the deposition process, facilitating the uniform and consistent application of perovskite precursor solutions onto substrates with meticulous accuracy, even across expansive surface areas. At its core, slot die coating embodies a delicate balance of artistry and engineering prowess. The process commences with the meticulous preparation of a perovskite precursor solution, a meticulous blend of lead halide salts, organic cations, and solvent mediums carefully calibrated to achieve the desired composition and properties conducive to optimal photovoltaic performance.

This approach, a solution-based and roll-to-roll compatible method, bears striking resemblances to the blade-coating technique while boasting a significant advantage of markedly decreased or zero material wastage, a common issue associated with blade-coating. Among various methodologies, this particular approach has demonstrated notable success in photovoltaic production. Referred to as the premiered coating process [8], it guarantees the creation of uniform, precisely regulated thickness films through meticulous control of material input. Its potential to transition from laboratory-based practices to large-scale industrial manufacturing in PSCs is particularly noteworthy [8,44,45]. When considering scalability and minimized material wastage, this process outperforms the spin-coating technique. The basic configuration of the slot die coater using a mini roll coater unit is illustrated in Figure 4.7.

German researchers have reportedly minimized ribbing effects in slot-die coating for manufacturing perovskite solar cells [8]. According to their claims, the achieved results represent the most efficient solar cell (>22% PCE) and mini-module of $12.7\,cm^2$, with a PCE of 17.1% constructed using this technique thus far.

4.3.4 Brush Painting

Brush painting emerges as a straightforward approach to fashioning perovskite solar cells, featuring a rapid fabrication process that minimizes material wastage. This method operates swiftly without requiring an annealing process, resulting in

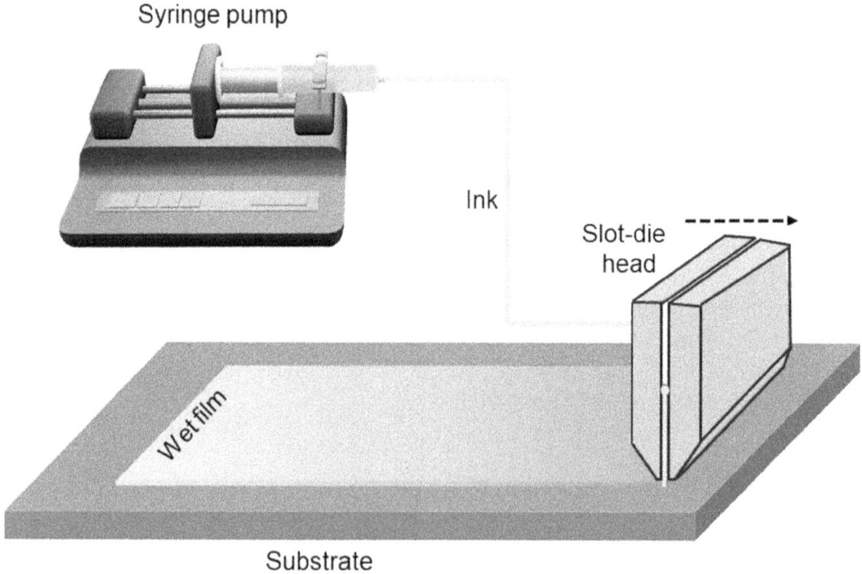

FIGURE 4.7 Slot-dye technique for perovskite film fabrication.

notably accelerated production. Its cost-effectiveness is pronounced, rendering it suitable for manufacturing large-area flexible solar cells. Achieving high-efficiency solar cells is feasible through meticulous optimization of solvent concentration and perovskite material. Additionally, this technique obviates the necessity for strict experimental conditions or time-consuming procedures [46]. Figure 4.8 showcases the construction of PSCs using the brush painting technique. Its advantages encompass operating sans vacuum and swift processing, contributing significantly to its cost-effectiveness. Devices crafted via brush painting exhibit improved efficiency

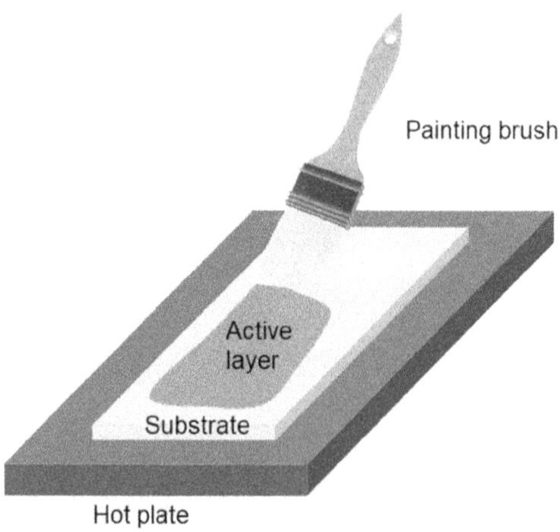

FIGURE 4.8 Brush painting technique for perovskite film fabrication.

compared to those produced through the spin-coating technique, given the absence of the need for thermal annealing. Nevertheless, the challenge of maintaining uniform thickness remains a limitation of brush painting. Fabricating the polymer solar cells using these techniques is easier compared to the PSCs due to ink properties and low boiling point solvent [47,48].

4.3.5 Electrodeposition

Electrodeposition (ED) of perovskite refers to the process of depositing perovskite materials onto a substrate using an electrochemical method. In this technique, a solution containing the precursor materials of the perovskite is prepared, and electrodes are immersed in this solution. By applying an electrical potential between the electrodes, ions from the solution are driven toward the electrodes, where they undergo chemical reactions to form the perovskite material. The perovskite then precipitates or forms a thin film on the electrode surface. This method offers advantages such as precise control over the deposition process, uniform film formation, and the ability to deposit perovskite materials on a variety of substrates. It is being explored as a promising approach for the fabrication of perovskite-based solar cells and other optoelectronic devices. Figure 4.9 illustrates the electrochemical deposition process of perovskite films.

In ED, a commonly employed setup involves a three-electrode system comprising a counter electrode (CE), a working electrode (WE), and a reference electrode (RE). These electrodes are submerged in an electrolyte solution, which can be reused multiple times, thus reducing production costs. The ED is a multistep process, in which first PbO [49] or PbI_2 [50] film is deposited. The perovskite films are achieved by exposing PbO or PbI_2 to the organic (MAI, FAI) cations] by ED process [51] or dipping process [52]. ED technique has the ability to deposit the perovskite layer on intricately shaped substrates with different device configurations, which makes it an attractive option for large-scale manufacturing [53].

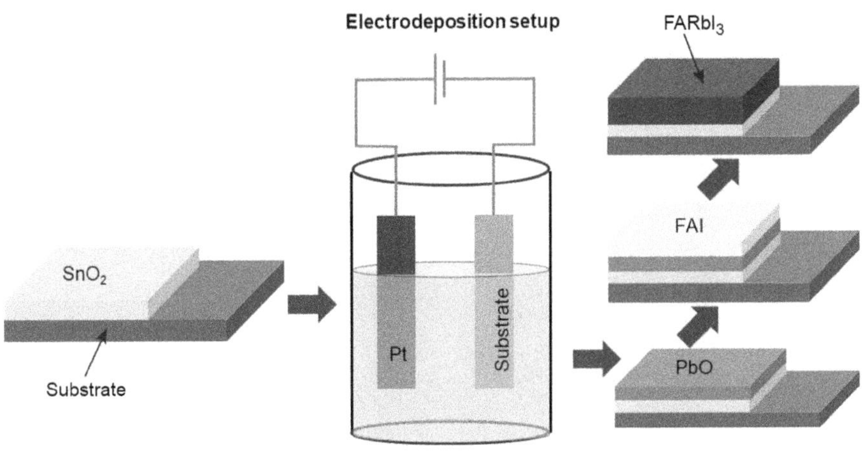

FIGURE 4.9 Depicts the schematic of PSC fabrication using the electrodeposition technique.

4.3.6 Ink-Jet Printing

The ink-jet printing represents a sophisticated and dynamic process, revolutionizing the deposition of organic, inorganic and perovskite materials onto a substrate through the intricate utilization of ink-jet printing technology. This technique has all the capability to fabricate the active layer and charge transport layer of OSC and PSCs on a large-scale [54]. Figure 4.10 illustrates the basic schematic of perovskite solar cell fabrication using the continuous and drop-on-demand ink-jet printing technique. Within this methodological marvel, a meticulously formulated solution brimming with the precursor materials essential for perovskite formation is meticulously loaded into the reservoir of an ink-jet printer cartridge. With precision akin to that of an artist's brushstroke, the printer orchestrates the dispensation of minuscule droplets of the solution onto the waiting substrate, navigating with unparalleled accuracy to fashion the desired pattern or layer. As these meticulously orchestrated droplets alight upon the substrate's surface, they embark upon a transformative journey, engaging in a mesmerizing dance of chemical reactions and processes. This intricate choreography ultimately culminates in the breathtaking emergence of the perovskite layer, a testament to the boundless potential harbored within each droplet. Ink-jet printing, with its unparalleled precision and finesse, not only offers a gateway to the realization of intricate designs but also boasts scalability, empowering the seamless replication of complex patterns across vast expanses. Moreover, its versatility transcends the limitations of traditional fabrication methods, affording researchers and engineers the freedom to explore new frontiers of design and functionality. As the relentless pursuit of innovation propels the quest for sustainable energy solutions ever forward, ink-jet printing emerges as a beacon of hope, heralding a new era in the large-scale

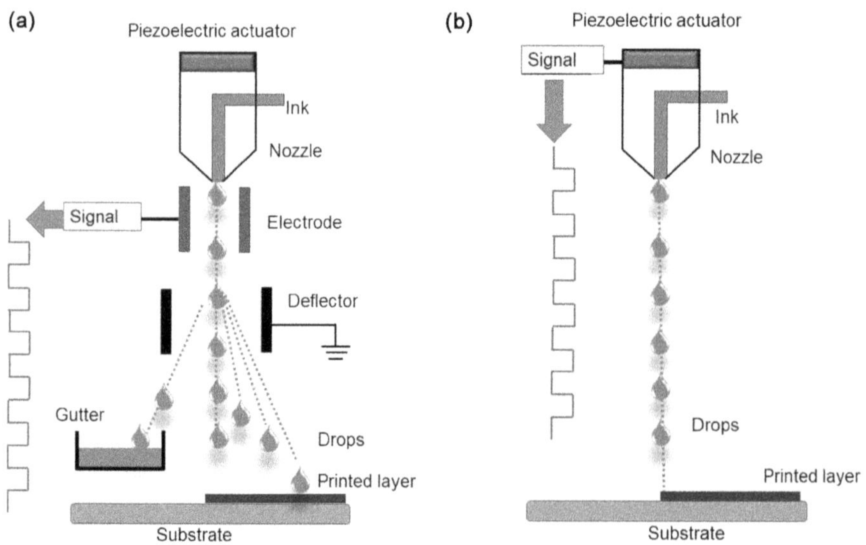

FIGURE 4.10 Schematic illustration of (a) continuous and (b) drop-on-demand ink-jet printing.

production of perovskite-based solar cells. Its promise of cost-effectiveness and high throughput beckons to the pioneers of tomorrow, inspiring them to embark upon a transformative journey toward a brighter, more sustainable future.

Key observations from the ink-jet printing technique encompass its exceptional precision in dispensing droplets of perovskite solution onto the substrate, enabling the creation of intricate patterns and layers with high fidelity. Additionally, this method offers scalability, allowing for the efficient production of perovskite-based devices on a large scale. Furthermore, ink-jet printing provides versatility, as it can accommodate various substrates and formulations, facilitating the exploration of novel materials and architectures. Another notable aspect is the potential for cost-effectiveness, as ink-jet printing minimizes material waste and can be easily integrated into existing manufacturing processes. Moreover, this technique enables rapid prototyping and customization, accelerating research and development efforts in the field of perovskite solar cells and other optoelectronic devices.

Generally, inkjet machines can be categorized as continuous mode inkjet (CIJ), and drop–on–demand (DOD) inkjet. In CIJ inkjet technology, a consistent flow of ink is produced by exerting pressure on an ink reservoir. This process is controlled by applying periodic stimulation through a piezoelectric actuator. When appropriately modulated, each nozzle can produce a continuous stream of hundreds of thousands of single droplets per second, each with a defined volume and velocity typically ranging from 10 to 20 mL/s. These droplets then traverse through an electric field where they are charged based on image information. As they progress downstream, they encounter high-voltage deflection plates, which direct them according to their charge. This ensures that they are deposited onto the correct spot on the substrate or redirected to an ink collection unit for recycling back to the feeding system. CIJ technology finds widespread application in high-volume, albeit relatively lower quality, fabrication tasks such as label printing, packaging, or personalized advertising applications. In Drop-on-Demand (DOD) technology, droplets are generated only when needed, providing precise control over the printing process. These printers can be further categorized based on their droplet ejection mechanisms. The deformation of the piezoelectric actuator creates a pressure wave, compelling a droplet to be expelled from the nozzle. DOD inkjet printing is more versatile for OSC and PSC applications because it completely satisfies the processing requirements. Modern printing machines reach throughput speeds of more than 1000 m hours.

Using inkjet printing perovskite layers with different composition perovskite precursors can be fabricated. Inkjet printed $Cs_{0.11}(FA_{0.83}MA_{0.17})_{0.89}$ $Pb(Br_{0.15}I_{0.85})_3$ perovskite layer atop NiO resulted in a PCE of 20.7% [55]. Also, $MAPbI_3$ and $CsFAMAPb(Br, I)_3$ perovskite layer atop TiO_2 have achieved maximum PCEs of 18.20% and 19.60% [56,57], SnO_2-based ETL and $Cs_{0.10}FA_{0.75}MA_{0.15}Pb(Br_{0.15}I_{0.85})_3$ have achieved PSC of 16.9% [58].

4.4 VAPOR-BASED TECHNIQUES

Solution-based and vapor-based methodologies stand out as the primary techniques for creating perovskite absorber layers, each offering distinct advantages and challenges. Among these, the vapor-assisted approach emerges as a particularly promising avenue due to its remarkable film uniformity and precise control over film thickness. Vapor-based techniques can be further categorized into physical and chemical methods, each with its own set of advantages and applications. Vapor deposition, a cornerstone of vapor-based techniques, enables the fabrication of highly crystalline films with nanometer-scale thicknesses, a stark contrast to the relatively thicker films produced by solution-based methods [59–62]. This nanoscale precision in film thickness is critical for optimizing light absorption and charge carrier transport within the solar cell device. Additionally, vapor-deposited thin films exhibit exceptional uniformity across large areas, offering superior performance compared to solution-processed films, which often suffer from non-uniformity and larger crystal grain sizes that can hinder device performance. In the realm of perovskite solar cell fabrication, the choice of deposition method profoundly impacts device performance. Vapor-based techniques, with their ability to produce precisely controlled, nanometer-thin films with high crystallinity, hold immense promise for achieving high-efficiency solar cells. By contrast, solution-based methods, while more accessible and cost-effective, often struggle to achieve the same level of uniformity and control over film properties.

4.4.1 Physical Vapor Deposition

Physical Vapor Deposition (PVD) of perovskite films is a process in which the precursor materials are evaporated in a vacuum chamber and then condensed onto a substrate to form thin films. Figure 4.11 illustrates the fundamental schematic diagram of PVD (co-deposition) for depositing perovskite films. This technique typically involves heating the precursor materials in a vacuum environment, causing them to vaporize and form a vapor cloud. The vaporized materials then travel through the vacuum chamber and deposit onto the substrate, where they condense and form a thin film. PVD offers several advantages for perovskite film deposition, including precise control over film thickness, high uniformity, and the ability to deposit films

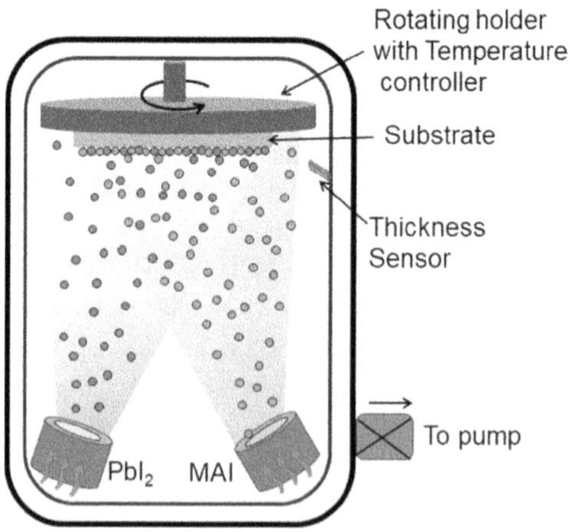

FIGURE 4.11 Schematic illustration of the vapor deposition technique utilized for fabricating perovskite films.

onto various substrates. By adjusting parameters such as temperature, pressure, and deposition rate, researchers can tailor the properties of the deposited films to meet specific requirements for perovskite solar cell applications.

Furthermore, PVD techniques can produce highly crystalline perovskite films with minimal defects, leading to improved device performance and stability. However, PVD methods may require specialized equipment and controlled environments, making them more challenging to implement on a large scale compared to solution-based techniques. Overall, physical vapor deposition of perovskite films holds promise as a versatile and effective method for fabricating high-quality thin films for use in perovskite solar cells and other optoelectronic devices. Continued research and development in this area are expected to further advance the efficiency and scalability of PVD techniques for perovskite film deposition.

Various thermal evaporation techniques were investigated for the deposition of perovskite layers, including co-evaporation, sequential evaporation, single-source evaporation, and hybrid vapor-solution deposition. Multisource co-evaporation stands as a commonly utilized technique for depositing perovskite absorbers. In this approach, both the metal halide (PbI_2, PbBr, CsI, SnI_2, SnF_2, etc.) and ammonium salt (MAI, FAI, etc.) are vaporized and deposited simultaneously. The process encompasses five fundamental steps: raw material sublimation, spatial transport, molecular absorption, and desorption, diffusion into the film, and precursor reaction and nucleation. Apart from co-deposition, magnetron sputtering [63,64], RF-sputtering [65], and pulsed laser deposition (PLD) [66–68] are also suitable for perovskite absorber deposition. Most recently, thermally evaporated perovskite films demonstrated highest PCE of 26.41% with a certified value of 26.21% [3].

4.4.2 Chemical Vapor Deposition (CVD)

Chemical Vapor Deposition (CVD) of perovskite films stands as a sophisticated process wherein precursor materials are meticulously introduced into a meticulously controlled reaction chamber, typically in the form of gases or vapors. Within the confines of the chamber, these precursor molecules undergo a meticulously orchestrated series of chemical reactions, ultimately culminating in their deposition onto a substrate to form a delicate yet robust thin film of perovskite material. In the realm of CVD, the substrate assumes a pivotal role, typically subjected to precise heating to a specific temperature [69,70]. This controlled thermal environment not only serves to facilitate the requisite chemical reactions but also exerts meticulous control over the growth of the perovskite film. Subsequently, the meticulously tailored precursor gases are introduced into the chamber, where they interact in a harmonious ballet of molecular transformations, ultimately depositing onto the awaiting substrate surface. Schematic Figure 4.12 shows the CVD process for fabricating perovskite films.

The appeal of CVD for perovskite film deposition lies in its myriad advantages. Foremost among these is the unparalleled precision it offers in controlling film thickness, uniformity, and morphology. Moreover, CVD enables the deposition of perovskite films onto a diverse array of substrates, expanding the horizons of potential applications. Furthermore, CVD holds the promise of producing highly crystalline perovskite films with minimal defects, thereby bolstering device performance

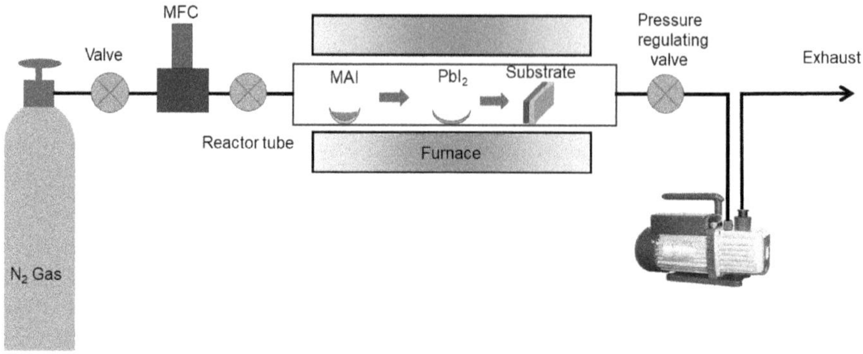

FIGURE 4.12 Schematics of the perovskite film fabrication using MAI and PbI_2 sources by CVD process.

and stability to unprecedented levels [71]. Nevertheless, it's important to acknowledge that CVD methodologies may necessitate specialized equipment and meticulously controlled environments. These requirements can pose challenges when it comes to implementing CVD techniques on a large scale, especially in comparison to more accessible solution-based approaches. Despite these challenges, CVD remains a cornerstone technique for fabricating high-quality perovskite films essential for advancing solar cells and other optoelectronic devices.

Chemical Vapor Deposition of perovskite films emerges as a versatile and effective method for meticulously crafting high-quality thin films endowed with precise control over properties crucial for optimal device performance. The relentless pursuit of research and development in this field is poised to further enhance the efficiency and scalability of CVD techniques for perovskite film deposition, propelling the realm of renewable energy and optoelectronics into new frontiers of possibility. In the CVD method, mostly organic halide precursors are used to deposition on PbI_2 to grow perovskite films. Tavakoli et al. used a single-step CVD process to fabricate $MAPbI_3$ perovskite film using PbI_2 and MAI sources yielded a PCE of 11.1% [72].

4.4.3 Vapor-Assisted Solution Process

The Vapor-Assisted Solution Process (VASP) for perovskite fabrication is a sophisticated technique that combines elements of both solution-based and vapor-based methodologies. Schematic Figure 4.13 shows fabricating perovskite films using the VASP technique. In VASP, a perovskite precursor solution is first deposited onto a substrate using conventional solution-based methods. Subsequently, a vapor, typically a solvent or anti-solvent, is introduced into the environment, facilitating the transformation of the precursor solution into a crystalline perovskite film [73–75]. The vapor serves multiple purposes in the VASP technique. Firstly, it aids in the nucleation and growth of perovskite crystals, promoting the formation of a dense and uniform film. Secondly, the vapor can help to control the crystallization kinetics and morphology of the perovskite film, leading to improved optoelectronic properties

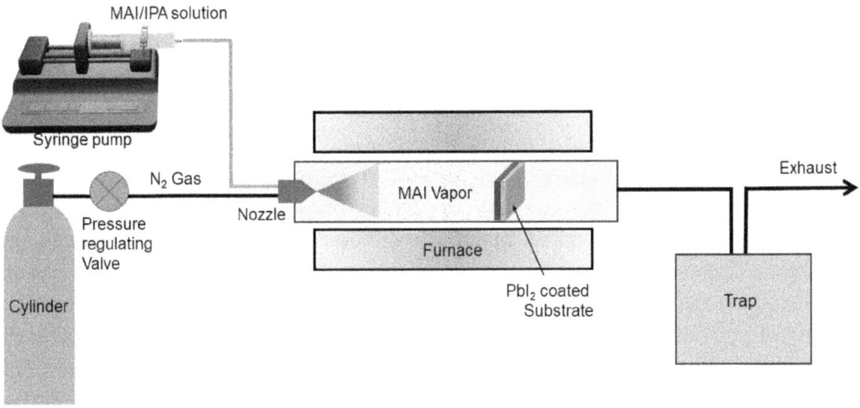

FIGURE 4.13 Schematic illustration of the VASP utilized for fabricating perovskite films.

and device performance. Additionally, the vapor can assist in the removal of excess solvent from the precursor solution, reducing defects and improving film quality. One of the key advantages of VASP is its ability to produce high-quality perovskite films with enhanced uniformity and crystallinity compared to traditional two-step solution-based methods [76,77]. By leveraging the benefits of both solution processing and vapor-phase deposition, VASP offers a versatile and scalable approach for fabricating perovskite films for a wide range of optoelectronic applications.

The vapor-assisted solution technique can be subdivided into several detailed fabrication processes: (a) Evaporation of the material onto the surface of the as-formed film. And (b) deposition of the film through the hot flow. Furthermore, the vapor-assisted solution method could serve as a post-treatment process aimed at reducing trap density and improving the quality of the perovskite film. Grätzel et al. fabricated $FAPbI_3$ perovskite films by spin coating and treated them with methylammonium thiocyanate vapors to improve the quality of the film and yielded a PCE of 23% [78].

4.5 PERFORMANCE, EFFICIENCY AND STABILITY AT THE LAB AND INDUSTRIAL LEVEL

The performance of perovskite solar cells has undergone an extraordinary evolution over the years, propelling them into the forefront as one of the most promising next-generation photovoltaic technologies. Key performance metrics of perovskite solar cells, including efficiency, stability, scalability, and cost-effectiveness, have witnessed unprecedented advancements. Here's an extended overview of the remarkable progress observed in the performance of perovskite solar cells:

Efficiency: Perovskite solar cells have achieved truly remarkable power conversion efficiencies (PCEs) in laboratory settings, with some single-junction cells surpassing an astonishing 25%. This remarkable surge in efficiency is attributed to relentless improvements in perovskite materials, innovative device architectures, intricate interface engineering, and cutting-edge fabrication techniques.

Researchers are vigorously pursuing the development of tandem and multijunction perovskite solar cells to further elevate efficiency levels and harness an even broader range of the solar spectrum.

Stability: Enhancing the stability of perovskite solar cells stands as a paramount objective for their widespread commercialization and practical deployment. Although perovskite materials are inherently susceptible to degradation when exposed to moisture, heat, and light, substantial research endeavors have been dedicated to fortifying stability through advanced material engineering, sophisticated encapsulation techniques, and meticulous interface optimization. While challenges persist, recent breakthroughs have resulted in significant strides towards enhancing the long-term stability of perovskite solar cells, bringing commercial viability within closer reach.

Scalability: Perovskite solar cells exhibit immense potential for scalable manufacturing owing to their versatile solution-based processing methods, seamless compatibility with roll-to-roll printing techniques, and utilization of cost-effective materials. Nonetheless, hurdles pertaining to material accessibility, refining manufacturing processes, and ensuring device stability must be surmounted to realize large-scale production and widespread commercialization of perovskite solar cells. Collaborative efforts across academia, industry, and government sectors are crucial for overcoming these challenges and accelerating the transition towards mass production and deployment.

Cost-effectiveness: Perovskite solar cells hold tremendous promise for achieving high cost-effectiveness due to their utilization of low-cost materials and streamlined solution-based processing techniques. As efficiency levels continue to ascend and manufacturing processes become increasingly optimized, the overall cost of perovskite solar cells is anticipated to decline substantially, further bolstering their competitiveness against conventional silicon-based photovoltaic technologies and driving widespread adoption across diverse applications. The performance of PSCs can be influenced by various factors, including material properties, device architecture, environmental conditions, and fabrication techniques. Here are some key factors affecting the performance of PSCs such as: Perovskite material quality, perovskite composition interfacial engineering, charge transport layers, device architecture, environmental stability fabrication techniques etc.

In the initial stages of PSC development, the prevailing perovskite composition was $MAPbI_3$ and these devices have attained a remarkable power conversion efficiency (PCE) exceeding 22% [79,80]. However, $MAPbI_3$'s practical application is hindered by its large band gap (1.57 eV), thermal instability, which encompasses volatility and intrinsic instability stemming from reversible/irreversible decomposition reactions, even at lower temperatures. Density functional theory (DFT) calculations show the octahedral tilting of $FAPbI_3$ is less pronounced compared to $MAPbI_3$ due to the FA cation has stronger hydrogen bonding with PbX_6 octahedra compared to the MA cation. As a result, $FAPbI_3$ exhibits a narrower bandgap of 1.43 eV and demonstrates improved thermal stability [81,82].

Furthermore, $FAPbI_3$ displays a reduced free volume, leading to a weaker electron-photon coupling and prolonged carrier lifetime [83,84]. Additionally, the calculated effective mass of $FAPbI_3$ is smaller than that of $MAPbI_3$, indicating excellent semiconducting characteristics [85]. Consequently, $FAPbI_3$ is endowed with superior carrier transport properties and an optimal bandgap, making it the most promising perovskite material for high-performance single-junction photovoltaic applications [86]. Nonetheless, α-$FAPbI_3$ represents a metastable phase at room temperature and is prone to transforming into the insulating δ phase. This means it is difficult to fabricate phase-pure α-$FAPbI_3$ films spontaneously. In general, $FAPbI_3$ possesses four polymorphs, as illustrated in Figure 4.14 [87,88]. At room temperature, α-$FAPbI_3$ demonstrates a cubic crystal structure, with the FA dynamically disordered within the lattice. α-$FAPbI_3$ transforms into the β phase upon cooling to 151 K. On further cooling to 91 K, the β phase transforms into the γ phase, with specific lattice tilting. Notably, the β and γ phases are only present at low temperatures so not suitable for PV applications. While δ-$FAPbI_3$, which is thermodynamically stable at room temperature features a hexagonal structure, and exhibits non-perovskite characteristics. The insulating properties and limited light absorption capability of δ-$FAPbI_3$ impede its potential for further applications in photovoltaics [89]. To address these critical problems, researchers have developed several techniques such as composition engineering, dimensionality engineering, strain engineering, and crystalline process modulation to stabilize α-$FAPbI_3$ phase and its application in PV devices.

In the compositional engineering methods [91,92] FA cation is partially replaced by MA, Cs, Rb etc. cations to stabilize the α-$FAPbI_3$ phase with enhanced optoelectronic and photovoltaic properties. Later on, these perovskites called double-cation FAMA or FACs, triple-cation CsMAFA, and quadruple-cation (rubidium-(Rb)-CsMAFA based perovskites) [93,94]. Among these strategies, methylammonium chloride (MACl) is commonly utilized as an additive to facilitate the high-quality crystallization of

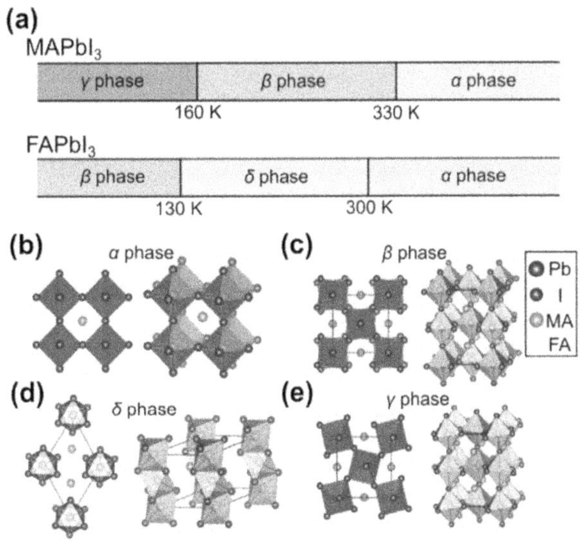

FIGURE 4.14 (a) The phase variations of $MAPbI_3$ and $FAPbI_3$ at various temperatures. There are four different phases: (b) the cubic alpha (α) phase, (c) the tetragonal beta (β) phase, (d) the trigonal delta (δ) phase, and (e) the orthorhombic gamma (γ) phase. (Reproduced from Ref. [90] under the CC BY-NC license.)

black-phase $FAPbI_3$ by forming an intermediate phase. This approach has successfully enhanced the efficiency of regular and inverted-structured devices. The efficiency of these PSCs utilizing FAPbI3-based quadruple cations or triple cations as the perovskite active layer exceeded 26% [2,95]. Consequently, compared to double cations or triple cations, $FAPbI_3$ faces challenges regarding the difficulty of fabricating perovskite films and the lower efficiency and stability of the final device, rendering it less competitive as the photoactive layer.

$FAPbI_3$ PSCs generally exhibit higher PCEs compared to $MAPbI_3$ PSCs. The narrower bandgap of FAPbI3 allows for better light absorption in the visible spectrum, leading to higher short-circuit current densities (J_{SC}) and overall higher efficiencies. $FAPbI_3$-based devices often achieve PCEs exceeding 26%, whereas $MAPbI_3$-based devices typically range from 20% to 23%. However, these values can vary depending on device architectures, fabrication techniques, and material optimizations. Figure 4.15 and Tables 4.1 and 4.2 present the annual efficiency progress of MAPbI3 and FAPbI3-based PSCs from the initial to now (March 2024).

The performance enhancement of $FAPbI_3$-based PSCs is more rapid comparted to the $MAPbI_3$-based PSCs due to extensive research and development efforts. Several major breakthroughs have been achieved in the development of $FAPbI_3$ PSCs, contributing to their rapid progress and improved performance. One significant breakthrough was the achievement of PCEs exceeding 20% by fabrication of high-quality $FAPbI_3$-based absorber layer using additive, compositional engineering, and charge carrier management [91,92,117]. In the case of n-i-p structured PSCs, SnO_2 ETL-based devices have performed much better than other ETL devices because they proved complete and uniform coverage, devoid of pinholes, ensuring optimal charge transport due to high electron mobility and good transparency in the visible spectrum. Additionally, it exhibits an ideal band alignment, facilitating efficient electron extraction from the perovskite layer while impeding hole back-transfer. SnO_2 facilitates efficient extraction and transport of electrons generated in the perovskite absorber layer, contributing to high device performance >25% [1,95,118]. For p-i-n architecture, PSCs organic HTL poly(triarylamine) (PTAA) and inorganic NiO have

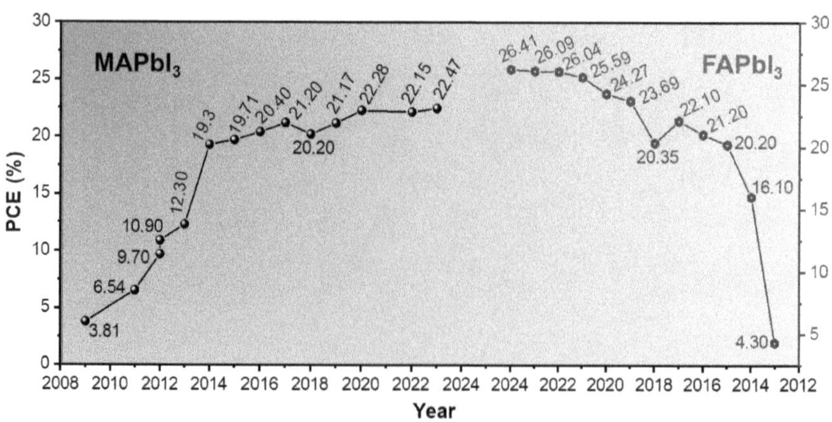

FIGURE 4.15 Development of $MAPbI_3$ and FAPbI3-based perovskite solar cells. (References are taken from Tables 4.1 and 4.2.)

TABLE 4.1
Annual Efficiency Progress of $MAPbI_3$ Based PSC Till March 2024

Year	Main Composition	J_{SC} (mA/cm²)	V_{OC} (V)	FF (%)	PCE (%)	References
2009	$MAPbI_3$	11.00	0.610	57.0	3.81	[96]
2011	$MAPbI_3$	15.82	0.706	58.6	6.54	[97]
2012	$MAPbI_3$	17.60	0.888	62.0	9.70	[98]
2012	$MAPbI_3$	17.80	0.980	63.0	10.90	[99]
2013	$MAPbI_3$	18.00	1.020	67.0	12.30	[100]
2014	$MAPbI_3$	22.75	1.130	75.1	19.30	[101]
2015	$MAPbI_3$	23.83	1.086	76.2	19.71	[102]
2016	$MAPbI_3$	23.69	1.113	77.3	20.40	[103]
2017	$MAPbI_3$	23.40	1.120	81.3	21.20	[104]
2018	$MAPbI_3$	22.70	1.100	81.0	20.20	[105]
2019	$MAPbI_3$	23.23	1.120	81.4	21.17	[106]
2020	$MAPbI_3$	24.11	1.159	79.7	22.28	[107]
2022	$MAPbI_3$	23.36	1.131	83.92	22.15	[79]
2023	$MAPbI_3$	22.48	1.29	77.49	22.47	[80]

TABLE 4.2
Annual Efficiency Progress of $FAPbI_3$ Based PSCs Till March 2024.

Year	Main Composition	J_{SC} (mA/cm²)	V_{OC} (V)	FF (%)	PCE (%)	References
2014	$FAPbI_3$	6.45	0.970	68.7	4.30	[108]
2014	$FAPbI_3$	20.97	1.032	74.0	16.01	[109]
2015	$FAPbI_3$	24.7	1.06	077.5	20.20	[110]
2016	$FAPbI_3$	23.70	1.134	78.0	21.02	[111]
2017	$FAPbI_3$	25.00	1.10	80.3	22.10	[112]
2018	$FAPbI_3$	25.06	1.080	75.5	20.35	[113]
2019	$FAPbI_3$	26.70	1.144	77.6	23.69	[114]
2020	$FAPbI_3$	26.18	1.181	79.6	24.27	[115]
2021	$FAPbI_3$	26.35	1.189	81.7	25.59	[116]
2022	$FAPbI_3$	26.66	1.193	81.31	26.04	[95]
2023	$FAPbI_3$	26.14	1.164	85.74	26.09	[2]

been widely explored in $FAPbI_3$ PSCs due to their suitable energy levels, high hole mobility, and good film-forming properties and shown similar PSC (>25%) as n-i-p structured devices [2,119]. Table 4.3 present the most efficient n-i-p and p-i-n structured $FAPbI_3$-based PSCs with certified PCE over 25%.

4.5.1 Stability of PSCs

$FAPbI_3$ perovskite materials have been mainly used in high-efficiency devices in recent years. The essential idea is to generate α-$FAPbI_3$ and avoid the generation of δ-$FAPbI_3$. A-site doping and additive engineering are the key techniques for the stabling of the perovskite **α-phase**. Since, FA^+ ion is too large, which makes the lattice

TABLE 4.3
The Summary of $FAPbI_3$ Based PSCs Certified PCE over 25% Reported in Literature

Year	Composition	Device Architecture	V_{OC} [V]	J_{SC} (mA/cm²)	FF (%)	Certified PCE (%)	References
2021	$FAPbI_3$	FTO/Cl-SnO_2/PVK/ Spiro-OMeTAD/Au	1.189	25.71	84.43	25.50	[120]
2021	$FAPbI_3$	FTO/c-TiO_2/m-TiO_2/PVK/ Spiro-OMe TAD/Au	1.189	26.35	81.70	25.20	[116]
2022	$FAPbI_3$	FTO/c-TiO_2/SnO_2/PVK/ Spiro-OMeTAD/Au	1.176	26.09	83.84	25.40	[121]
2022	$FAPbI_3$	FTO/SnO_2/PVK/ Spiro-OMeTAD/Au	1.182	26.30	82.70	25.60	[122]
2022	$Rb_{0.05}Cs_{0.05}MA_{0.05}FA_{0.85}$ $Pb(I_{0.95}Br_{0.05})_3$	ITO/MeO-2PACZ/PVK/ LiF/C_{60}/BCP/Ag	1.16	26.096	83.82	25.37	[123]
2022	$(FA_{0.98}MA_{0.02})_{0.95}Pb(I_{0.9}$ $_8Br_{0.02})_3$	ITO/PTAA/PVK/$FcTc_2$/ C_{60}/BCP/Ag	1.21	24.90	84.65	25.39	[124]
2023	$FAPbI_3$	FTO/SnO_2/PVK/ Spiro-OMeTAD/Au	1.191	26.39	82.94	25.80	[95]
2023	$FAPbI_3$	FTO/SnO_2/PVK/ Spiro-OMeTAD/Au	1.178	25.69	86.15	25.73	[1]
2023	$Cs_{0.05}MA_{0.05}FA_{0.9}PbI_3$	FTO/TiO_2/SnO_2/PVK/ Spiro-OMeTAD/Au	1.16	26.30	83.0	25.02	[125]
2022	$FAMAPb(I,Br)_3$	FTO/SnO_2/PVK/ Spiro-MeOTAD/Ag	1.164	26.14	85.74	25.81	[126]
2022	$(FAPbI_3)_{0.99}(MAPbBr_3)_{0.01}$	ITO/SnO_2/PVK/ Spiro-OMeTAD/Au	1.175	25.52	83.88	25.0	[127]
2023	$(FAPbI_3)_{0.99}(MAPbBr_3)_{0.01}$	FTO/SnO_2/PVK/ Spiro-OMeTAD/Au	1.17	26.12	85.22	25.06	[118]
2023	$Cs_{0.05}MA_{0.05}FA_{0.9}PbI_3$	FTO/TiO_2/SnO_2/PVK/ Spiro-OMeTAD/Au	1.159	26.01	83.90	25.04	[128]
2023	$Cs_{0.05}FA_{0.85}$ $MA_{0.1}PbI_3$	ITO/NiO_x/MeO-4PADBC/ PVK/C_{60}/BCP/Ag	1.190	25.40	84.60	25.58	[119]
2023	$Cs_{0.05}(FA_{0.95}MA_{0.05})_{0.95}$ $Pb(I_{0.95}Br_{0.05})_3$	ITO/MPACPA/PVK/C_{60}/ BCP/Ag	1.21	24.78	84.65	25.40	[129]
2023	$FA_{0.95}Cs_{0.05}PbI_3$	ITO/PTAA/PVK e/C_{60}/ BCP/Ag	1.153	26.14	84.46	25.81	[2]
2023	$CsFAMAPb(I,Br)_3$	ITO/DMAcPA/PVK/ $PC_{61}BM$/BCP/Ag	1.187	25.69	84.73	25.39	[130]
2024	$FA_xMA_{x-1}PbI_3$	FTO/SnO_2/PVK/ Spiro-OMeTAD/Au	1.166	26.169	82.474	25.165	[131]
2024	$FAPbI_3$	FTO/SnO_2/PVK/ Spiro-OMeTAD/Au	1.175	26.47	84.94	26.21	[3]

unstable, the researchers consider using some cations such as MA^+, Cs^+, and Rb^+ with a smaller ion radius to partially replace FA^+ [91,132,133]. In this way, the tolerance factor can be effectively adjusted to maintain the stability of the α-phase. Apart from this, several organic and inorganic additives have been reported in the literature that help to stabilize the **α-$FAPbI_3$** [134–136].

The stability of PSCs is affected by various factors, including aging conditions, perovskite composition, ETL/HTL, passivation, encapsulation, and electrode materials, making it a complex challenge. There is various factors that are considered for the stability of PSC such as:

- Inherent stability of perovskite phase
- Impact of environment (moisture and O_2) on stability
- Perovskite degradation under illumination
- Perovskite degradation under thermal stress
- Impact of defects on PSC stability
- Perovskite degradation under bias
- Impact of charge-transport layers on PSC stability
- Impact of metal electrodes on PSC stability

The inherent stability of any perovskite is determined by the Goldschmidt tolerance factor τ:

$$\tau = (R_A + R_X) / 2^{1/2}(R_B + R_X)$$

where R_A, R_B, and R_X are the ionic radii of the corresponding ions.

In general, if τ is between 0.9 and 1.0, it forms an ideal cubic structure. If τ is between 0.71 and 0.9, it will form an orthorhombic structure at room temperature. Observations have shown that when the parameter τ exceeds 1 or falls below 0.8, the lattice distortion induced leads to the formation of a photo-inactive, non-perovskite structure, commonly referred to as the δ-phase [137].

$MAPbI_3$, with a τ value of ~0.9, exhibits an ideal cubic structure at room temperature. While, for $FAPbI_3$, $\tau \sim 1$, consequently it exhibits two phases at room temperature: cubic (α-phase) and hexagonal (δ-phase). Figure 4.14 gives a clear idea of the phase transformation of $FAPbI_3$ with temperature. Stabilizing cubic $FAPbI_3$ is challenging due to its higher τ value. Another inorganic perovskite, $CsPbI_3$, with a τ value of around 0.8, lies close to the lower range for the cubic perovskite structure, gives instability it at room temperature, and transforms to photo-inactive δ-phase. At high temperature >633 K, stable photo-active α-$CsPbI_3$ can be obtained. At room temperature stable γ-$CsPbI_3$ can be obtained by additive and compositional engineering [138,139].

Mostly all, perovskite materials are sensitive to moisture and humidity, which can lead to degradation through processes such as hydrolysis and ion migration. Exposure to moisture can cause the formation of lead halide hydrates, which destabilize the perovskite structure. Degradation of perovskite under illumination is also a challenge for its long-term stability. The light illumination causes the formation of reactive species, intermediate species, and ions that alter the perovskite composition and stability. These defects and intrinsic defects in perovskite cause charge carrier trapping that causes bimolecular reactions. This reaction further generates the halide (I, Cl, Br) defect in the perovskite. These halides are mobile under the voltage bias within the perovskite and accommodate at the interfaces causing the degradation of charge transport layer as well as electrodes [140].

Numerous strategies have been devised to enhance the stability of perovskite-based devices. These include:

Alloying to increase crystal stability: Introducing additional elements through alloying can increase the stability of perovskite crystals by raising the energy barrier for ion migration and improving crystal structure robustness. Table 4.4 shows some stability data of PSCs in which perovskite is stabilized by A-cation allying.

Protection from external stressors: Implementing protective measures to shield both the perovskite layer and charge-transport layers (CTLs) from external stressors such as moisture, oxygen, light, and mechanical strain helps to preserve their structural integrity.

TABLE 4.4
Summary of PSC Stability Enhancements from A-Cation Alloying

Perovskite	PCE (%)	Stability Performance-Conditions	Stability Results	References
$Rb_x(Cs_5MA_{0.17}FA_{0.83}Pb(I_{0.83}Br_{0.17})_3)_{100-x}$	21.6	MPPT, 85°C, N_2 atmosphere, 1-sun continuous illumination, 500 hours	Decay 5%	[133]
$DMA_{0.1}FA_{0.6}Cs_{0.3}PbI_{2.4}Br_{0.6}$	18	85°C in the dark, open circuit, N_2 atmosphere, 2200 hours	Maintain 95%	[141]
$MA_{0.1}FA_{0.75}Cs_{0.15}PbI_3$	20.44	Storage, RT, RH 40%–60%, 1000 hours	Maintain ~80%	[142]
$(MA_{0.7}FA_{0.3})_{0.97}EDA_{0.015}PbI_3$	20.01	Storage, ~27°C, RH ~ 57%, 25 days	Retain ~96%	[143]
$(FA_{0.8}MA_{0.15}Cs_{0.05})_{0.85}HA_{0.15}Pb(I_{0.85}Br_{0.15})_3$	21.20	Storage, 25°C ± 5°C, RH 30% ± 5%, 2520 hours	Maintain ≈90%	[144]
$FABACsPbI_3$	22.5	Air exposure, RH 60%–65%, 2 months	Retain ~95%	[145]
$MA_{0.17}FA_{0.83}Pb(I_{0.83}Br_{0.17})_3$	21.1	MPPT, 1-Sun, RT, N_2 atmosphere, 250 hours	PCE drop from 16% to 8% in 100 hours	[93]
$Na_{0.05}(FA_{0.85}MA_{0.15})_{0.95}Pb(I_{0.85}Br_{0.15})_3$	20	Storage, RT, ~20 RH%, 1000 hours	Maintain ~95%	[146]
$MA_{0.75}Gua_{0.25}PbI_3$	19	MPPT, AM1.5G Sun illumination, 60°C, Ar atmosphere, 1100 hours	Maintain ~80%	[147]
$Rb_{0.05}Cs_{0.05}MA_{0.05}FA_{0.85}Pb[I_{0.95}Br_{0.05}]_3$	25.0	Continuous illumination (100 mW/cm^2), 2400 hours	Maintained ≈87%	[123]
$Cs_{0.05}(FA_{0.98}MA_{0.02})_{0.95}Pb(I_{0.98}Br_{0.02})_3$	25.0	MPPT 1500 hours at 1 Sun illumination	Maintained ≈98%	[124]
$FAPbI_3$:0.38$MDACl_2$	25.5	500 hours MPP(N_2)	Maintain 90%	[120]
$(FAPbI_3)_{1-x}(MACl)_x$	25.7	350 hours MPP(N_2)	Maintain 95%	[121]

Passivation of interfacial defects: Applying barrier layers or passivation coatings effectively seals interfacial defects, preventing detrimental interactions between the perovskite layer and adjacent layers, thereby enhancing device stability.

Reduction of defect density: Incorporating functional additives can lower the density of defects within the perovskite layer, thereby reducing non-radiative recombination and enhancing device stability.

Suppression of ion migration: Certain additives or treatments can suppress ion migration within the perovskite layer, mitigating structural degradation and improving long-term stability.

Utilization of robust CTLs: Employing robust charge-transport layers with enhanced chemical and thermal stability can contribute significantly to the overall stability of perovskite-based devices.

4.5.1 Dimensional Engineering in PSCs

The incorporation of bulky organic cations (BOCs) can directly impact the crystalline growth of perovskite, resulting in modifications to its dimensionality, optoelectronic properties, performance, and stability. Perovskite materials can be categorized into different dimensional types, including 2D, 3D, and quasi-2D, depending on the number of inorganic slabs (*n*) within their crystal structures. For instance, a purely 2D perovskite has an *n* value of 1, whereas quasi-2D perovskites demonstrate *n* values ranging from 1 to 60 ($1 < n \leq 60$), as illustrated in Figure 4.16. Higley-oriented 2D perovskites are of different types, such as the Ruddlesden-Popper (RP) phase, Dion-Jacobson (DJ) phase, and the phase with alternating cations in the interlayer space (ACI), depending on the type of BOCs used [148]. The formation of 2D perovskites is notably preferred when employing BOCs with an ionic radius exceeding the critical size recommended by the Goldschmidt tolerance factor. BOCs commonly utilized for this purpose encompass ammonium, amidinium, phosphonium, imidazolium, pyridinium, and piperidinium [148–150].

Several BOCs, such as PEAI, OAI, FPEAI, TMAI, etc., have been used in passivation treatments to form 3D/2D bilayer or heterophase perovskite with high

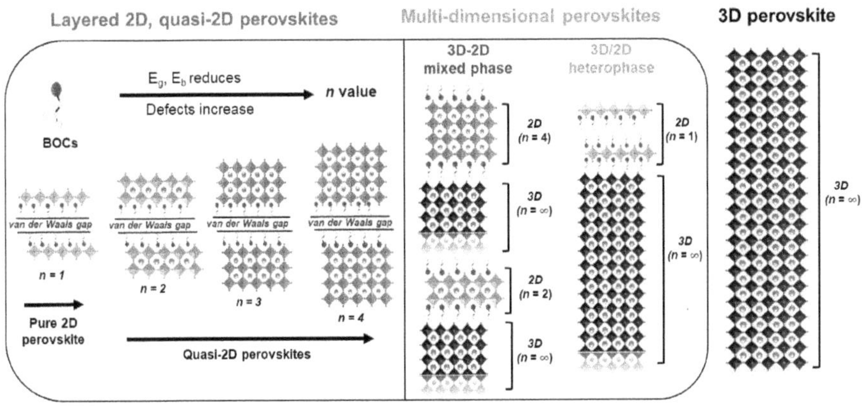

FIGURE 4.16 Schematic figure illustrating layer perovskites with different n values and dimensionalities. (Adapted with permission from Ref. [148] Copyright 2021, Elsevier.)

efficiency and stability. The effect of BOCs on device performance and stability is clearly reported in a review article by Lee et al. [148] and Metcalf et al. [151]. The BOC treatment on 3D perovskite enhances the perovskite phase stability and operational stability of PSC and modules. Table 4.5 presents the interface engineered PSCs mini-modules reported in the literature. In the year 2024, $FA_{0.3}MA_{0.7}PbI_3$ perovskite passivated with 4-F-PEAI+nDDAI yielded the PCE of a mini-module up to 19.2% at an active area of 108 cm² Figure 4.17a–c [152]. Most recently, $Cs_{0.05}MA_{0.05}FA_{0.90}PbI_3$ perovskite with high crystallinity and less defects caused by the passivation of PEAI have shown the fabrication of perovskite solar modules with a certified efficiency of 23.30% and ultimately stabilized at 22.97% over a 27.22 cm² aperture area, marking the highest certified PSM performance [9]. These modules were manufactured on rigid substrates such as ITO or FTOs. The fabrication process for PSCs on individual sheets involves vacuum-based evaporation steps and employs subtractive laser patterning to establish interconnections for large-area modules. These requirements pose challenges in reducing the cost of large-scale production. In contrast, flexible PSCs enable high-volume and efficient manufacturing through continuous roll-to-roll (R2R) techniques such as slot die, reverse gravure, and screen printing as shown in

TABLE 4.5
Interface Engineering for the Mini-modules of PSCs

Perovskite	Materials	Active Area (cm²)	PCE (%)	Stability	Year	References
$Cs_{0.05}FA_{0.54}MA_{0.41}Pb(I_{0.98}Br_{0.02})_3$	EDTAK	22.4	16.6	86% (2000 hours, MPPT)	2020	[154,155]
$MAPbI_3$	5-AVAI	70	11.55	60% (2700 hours, RH 15%–20%)	2021	[153]
$Cs(FAPbI_3)_x(MAPbBr_3)_{1-x}$	*o*-PDEAI$_2$	26	21.36	90% (1100 hours, MPP, N_2)	2021	[156]
$FA_xCs_{1-x}PbI_3$	FPEAI	30	18.6	93.6%, >1000 hours operational	2021	[157]
$FA_{0.83}Cs_{0.17}PbI_3$	KPF_6	17.0	20.4	NIM China certified	2021	[158]
$MA_{0.6}FA_{0.4}PbI_3$	*p*-F-PEAI+nDDAI	18.1 50.0	19.25 19.15	NREL certified	2022	[159]
$(FAPbI_3)_{0.95}(MAPbBr_3)_{0.05}$	DIC-PBA	33.2	19.13	90% (1200 hours, ambient condition)	2022	[160]
$CsFAMAPb(I_{1-x}Br_x)_3$	CPPA	17.1	21.4	95.5%, 3265 hours MPPT	2023	[161]
$(FAPbI_3)_{0.95}(MAPbBr_3)_{0.05}$	4F-PEAI	17.1	19.60	90%, 4390 hours MPPT	2023	[162]
$Cs_{0.05}MA_{0.05}FA_{0.9}PbI_3$	BABr and BAI	29.0	21.48	90%, 2000 hours MPPT	2023	[128]
$FA_{0.3}MA_{0.7}PbI_3$	4-F-PEAI+nDDAI	78 84 108	19.47 19.55 19.21	NREL stabilized efficiency	2024	[152]
$Cs_{0.05}MA_{0.05}FA_{0.90}PbI_3$	PEAI	27.22	22.97	NPVM certified	2024	[9]

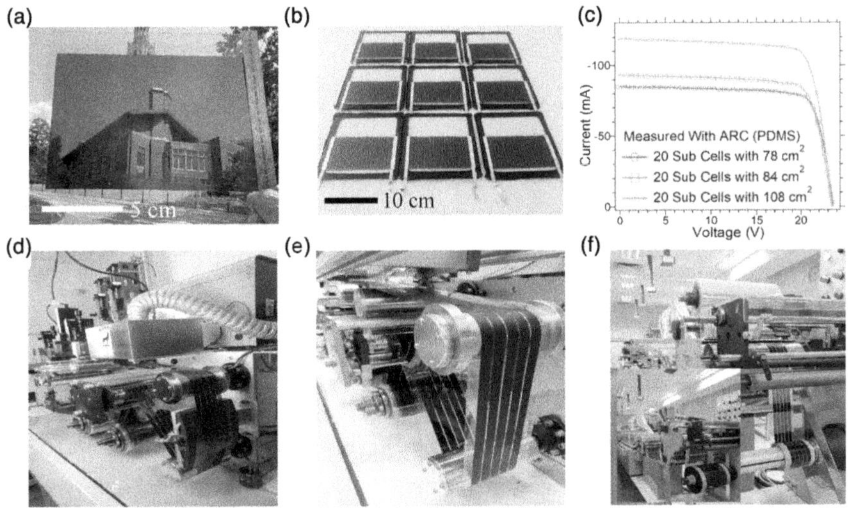

FIGURE 4.17 (a) A photograph of a blade-coated $FA_{0.3}MA_{0.7}PbI_3$ perovskite film on an ITO substrate with an area of ~130 cm^2, (b) a photograph of encapsulated minimodules with aperture areas of ~78 cm^2, (c) *J–V* curves of the minimodule with aperture areas of 78, 84, and 108 cm^2, respectively. (Adapted with permission from Ref. [152] Copyright 2021, Elsevier.) (d) Image of a 5-channel slot-die coater. (e) Reverse gravure (RG) coating of carbon film on the TCE/ETL/perovskite/HTL stack (f) Image of the R2R-screen-printed perovskite PV modules. (Reproduced from Ref. [90] under the CC BY-4 DEED license.)

Figure 4.17d–f. The lightweight and flexibility of flexible PSCs also offer the potential for solar PV panels with high specific power (power-to-weight ratio), which is particularly beneficial for emerging applications such as space, vehicle-integrated PV, and building-integrated PV [153]. The flexible PSC large-area module (50 cm^2) fabricated by these techniques demonstrated a PCE of 11.0%. The Australian group has reported the production of flexible PSC modules at the cost of 0.7 USD/watt for a production rate of 1,000,000 m^2/year [153]. This work signifies a notable advancement in perovskite solar technology toward achieving low-cost, large-scale commercial manufacturing capabilities.

4.6 ORGANIC SOLAR CELLS AND THEIR PROGRESS

The initial configuration of OSCs was a single active layer positioned between electrodes with different work functions. Nevertheless, these devices with only one layer showed poor PCE, remaining below 0.1%. This was primarily attributed to difficulties in efficiently segregating excitons and encountering notable electron-hole recombination. A significant milestone occurred in 1986 when Tang introduced the concept of a bilayer heterojunction structure, which greatly improved OSC efficiency [163]. This configuration incorporated copper phthalocyanine as the donor (D) and a perylene tetracarboxylic derivative as the acceptor (A). This breakthrough marked a significant advancement in the field of OSCs. Utilizing copper phthalocyanine and perylene

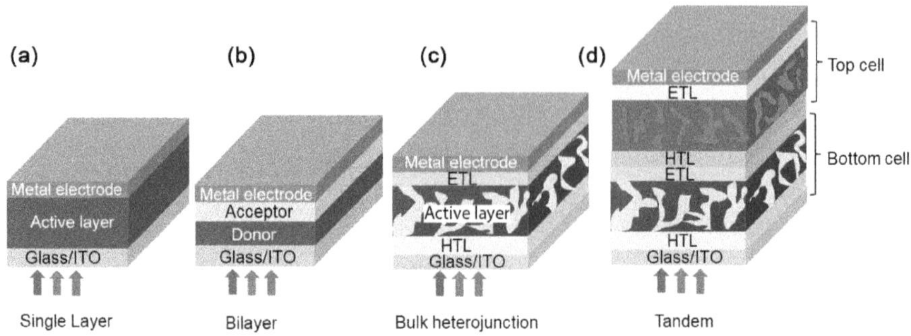

FIGURE 4.18 Different configurations of OPV devices (a) single layer, (b) bilayer, (C) bulk-heterojunction, and (dD) tandem structure.

tetracarboxylic derivative as active layers within this bilayer heterojunction setup yielded an impressive PCE of approximately 1%. However, the limited interface area between the donor and acceptor persisted as a barrier to efficient exciton diffusion and separation, preventing bilayer OSCs from achieving high PC [163]. In 1995, bulk heterojunction (BHJ) OSCs were introduced for the first time (illustrated in Figure 4.18). This architecture involved blending electron-donating and electron-accepting materials at the nanoscale to maximize the donor-acceptor interface and enhance charge separation. The BHJ structure's introduction led to substantial improvements in the donor/acceptor interface and reduced the distance for exciton separation, resulting in remarkable enhancements in device performance. Since its inception, the BHJ structure, hailed as a breakthrough in OSCs, has witnessed a significant surge in PCE to over 19% establishing it as a promising PV technology [164,165].

4.6.1 Device Structure of OPV Cells

4.6.1.1 Single-Layered OPV Cells

These represent the most fundamental and rudimentary form of organic OPV cells, embodying a singular layer composed solely of organic materials meticulously engineered to adeptly capture photons and catalyze the generation of electrical energy. Single-layered OPV cells typically manifest diminished efficacy when juxtaposed with their counterparts of more intricate OPV cell configurations. Within the confines of single-layer OPV cells, the donor and acceptor materials harmoniously intertwine within the very same stratum, thereby engendering a streamlined and uncomplicated device architecture, potentially eliciting a reduction in manufacturing expenses. Upon the illumination of the donor-acceptor layer by incident light, a cascade of events ensues whereby photon absorption catalyzes the generation of electron-hole pairs, which, in turn, undergo spatial segregation propelled by the inherent electric field. The ensuing partitioning of electrons and holes serves as a prelude to their respective captivation by the electrodes, thereby instigating the initiation of an electric current—a fundamental phenomenon that furnishes the capability to energize external apparatuses or alternatively stockpile the generated electrical energy within a storage medium such as a battery. Figure 4.18 and Table 4.6 present

TABLE 4.6
A Brief Comparison between Device Structures of OPV Cells

Structure	Advantages	Disadvantages	Key Features
Single-layered	Simple device architecture and potentially lower production cost	Lower efficiency compared to other OPV cells	A single layer of organic material that combines both the donor and acceptor materials, three layers in total
Bilayer	Higher efficiency compared to single-layered OPV cells	More complex device architecture than single-layered OPV cells	Two layers of organic semiconductor sandwiched between two electrodes, solution-based methods such as spin-coating
Bulk heterojunction (binary)	High efficiency and excellent flexibility	Complexity in blending donor and acceptor materials	Interpenetrating network of electron-donating and electron-accepting materials, typically deposited by spin-coating
Bulk heterojunction (ternary)	High efficiency	Complexity in blending donor and acceptor materials	Expanded absorption spectrum and improved exciton dissociation, enhanced morphology, tailored energy levels
Tandem	Higher efficiency due to a wider range of solar spectrum	More complex device architecture than single-layered OPV cells	Two or more sub-cells connected in series, with different band gaps to absorb different parts of the solar spectrum

a comparison of the device structures of OPV cells, along with their advantages, disadvantages, and key features.

The single-layer OPV cell typically consists of the following layers:

Transparent electrode: This layer is typically made of a transparent conductive oxide material, such as indium tin oxide (ITO) or zinc oxide (ZnO), and is used as the anode.

Donor–acceptor blend active layer: This layer is a single layer that combines both the donor and acceptor materials. The donor material is typically a polymer or small molecule that absorbs light and generates electrons, while the acceptor material is typically a fullerene derivative that accepts electrons.

Interfacial layers: In some designs, additional layers may be included to improve the interface between the active layer and electrodes, enhancing charge extraction and reducing losses due to recombination. Examples include holes or electron transport layers.

Metal electrodes: These layers serve as contacts for extracting electrical current generated by the cell. Typically, a transparent conductive layer (such as indium tin oxide, ITO) is used as the bottom electrode, while a metal layer (such as aluminum or silver) is used as the top electrode.

4.6.1.2 Bilayer OPV Cells

Bilayer OPV cells represent a more advanced configuration compared to their single-layer counterparts. In bilayer OPV cells, two distinct layers of organic materials are strategically arranged to optimize light absorption and charge separation, thereby enhancing device efficiency. Typically, one layer serves as the donor material, responsible for absorbing photons and generating electron-hole pairs upon illumination, while the other layer acts as the acceptor material, facilitating the efficient separation and transport of charge carriers. The bilayer structure allows for a more precise control of the donor-acceptor interface, which is crucial for maximizing the efficiency of charge generation and collection. This configuration also offers the flexibility to tailor the optical and electronic properties of each layer independently, thereby enabling optimization of the device for specific application requirements. Moreover, bilayer OPV cells often exhibit improved stability and reliability compared to single-layer cells, as the separate layers can be engineered to mitigate degradation mechanisms and enhance long-term performance. Overall, bilayer OPV cells represent a significant advancement in organic photovoltaic technology, offering higher efficiency, improved stability, and greater versatility for various applications in renewable energy generation.

4.6.1.3 Bulk Heterojunction OPV Cells

Bulk heterojunction (BHJ) OPV cells represent a cutting-edge design that has revolutionized the field of organic solar cells. Unlike traditional single-layer or bilayer configurations, BHJ OPV cells feature a finely intermixed blend of donor and acceptor materials at the nanoscale level throughout the active layer. This intimate mixing of donor and acceptor materials creates a large interfacial area, facilitating efficient exciton dissociation and charge carrier generation. The disordered morphology of the BHJ structure promotes multiple exciton dissociation pathways and efficient charge transport, leading to enhanced device performance. The bulk heterojunction architecture enables versatile material design, allowing researchers to tailor the energy levels, absorption spectra, and charge transport properties of the donor and acceptor materials to optimize device performance. This flexibility has led to significant improvements in PCE and stability compared to traditional OPV designs. BHJ OPV cells have become the focus of extensive research and development efforts due to their potential for low-cost, scalable manufacturing and compatibility with flexible substrates. They hold promise for a wide range of applications, including portable electronics, building-integrated photovoltaics, and off-grid power generation. Overall, bulk heterojunction OPV cells represent a promising avenue for advancing renewable energy technologies and addressing global energy challenges. Continued research and innovation in materials science, device engineering, and manufacturing processes are expected to further improve the performance and commercial viability of BHJ OPV technology.

4.6.1.4 Tandem OPV Cells

Tandem organic photovoltaic (OPV) cells are an innovative design that aims to further improve the efficiency of solar energy conversion. In tandem OPV cells, two or more individual solar cell layers are stacked on top of each other, with each layer optimized to absorb a specific portion of the solar spectrum. The tandem architecture allows for more efficient utilization of sunlight by enabling each cell layer to absorb different wavelengths of light. This approach mitigates losses due to incomplete absorption and thermalization of photons, leading to higher overall power conversion efficiency. Typically, tandem OPV cells consist of a combination of different active materials with complementary absorption spectra. For example, a tandem cell may feature a top cell optimized for absorbing short-wavelength photons (e.g., blue and green light) and a bottom cell tailored to absorb longer-wavelength photons (e.g., red and near-infrared light). This division of labor maximizes light harvesting across the entire solar spectrum. Moreover, tandem OPV cells can be engineered to minimize current matching issues between the individual cell layers, thereby enhancing overall device performance. Various strategies, such as interconnecting layers with transparent conductive layers or incorporating intermediate layers to adjust the optical properties, can be employed to optimize tandem cell performance. Tandem OPV cells hold great promise for achieving high-efficiency solar energy conversion while maintaining the advantages of organic materials, such as flexibility, lightweight, and low-cost manufacturing. Continued research and development efforts in tandem cell design, materials optimization, and fabrication techniques are expected to further advance the efficiency and scalability of tandem OPV technology, paving the way for its widespread adoption in renewable energy applications.

Organic semiconducting materials play a crucial role in the development of organic photovoltaic (OPV) cells, serving as the active components responsible for converting sunlight into electrical energy. These materials possess unique electronic properties that make them well suited for use in solar cells. Some of the key organic semiconducting materials utilized in OPV include:

Conjugated polymers: Conjugated polymers, also known as π-conjugated polymers, are long chains of repeating units with alternating single and multiple bonds, resulting in delocalized π-electrons along the polymer backbone. Examples include polythiophene, poly(3-hexylthiophene) (P3HT), and poly(3,3″-didodecylquaterthio phene) (PQT-12). Conjugated polymers offer excellent processability, allowing for solution-based fabrication methods such as spin-coating and printing.

Fullerene derivatives: Fullerene derivatives, particularly [6,6]-phenyl-C_{61}-butyric acid methyl ester (PCBM), are widely used as electron acceptors in OPV cells. These spherical carbon molecules possess high electron affinity and can efficiently accept and transport electrons generated upon photon absorption.

Non-fullerene acceptors (NFAs): Non-fullerene acceptors have emerged as promising alternatives to fullerene derivatives due to their tunable energy levels, strong absorption in the visible spectrum, and superior charge transport properties. NFAs include various classes of organic molecules such as small molecules (e.g., ITIC, IT-4F), oligomers, and polymers (e.g., PTB7-Th).

Small molecule donors: Small molecule organic semiconductors, such as pentacene and perylene diimide derivatives, exhibit well-defined molecular structures and high crystallinity, leading to efficient charge transport. These materials are often used as donor components in OPV cells, either alone or in combination with fullerene or non-fullerene acceptors.

Interfacial materials: Organic materials used at the interfaces between different layers in OPV cells play a crucial role in charge extraction, carrier transport, and device stability. Examples include hole transport materials (e.g., PEDOT:PSS), electron transport materials (e.g., PCBM), and interfacial modifiers (e.g., self-assembled monolayers).

The selection and optimization of organic semiconducting materials are essential for achieving high-performance OPV cells. Ongoing research efforts focus on developing novel materials with improved absorption, charge transport, stability, and scalability, aiming to further enhance the efficiency and commercial viability of organic solar technology.

Beyond the conventional configuration depicted in Figure 4.18, there exist more advanced configurations of OSCs, such as the ternary configuration illustrated in Figure 4.19. In ternary blend systems for OPV devices, the combinations of three different materials—two donor materials and one acceptor material (D1:D2:A), or one donor and two acceptors (D:A1:A2)—offer a vast array of possibilities for optimizing device performance [166,167]. Ternary solar cells encompass three fundamental concepts: charge transfer, energy transfer, and the parallel-linkage or alloy structure, all of which are substantially affected by the introduction of the third component, whether it acts as the second donor or

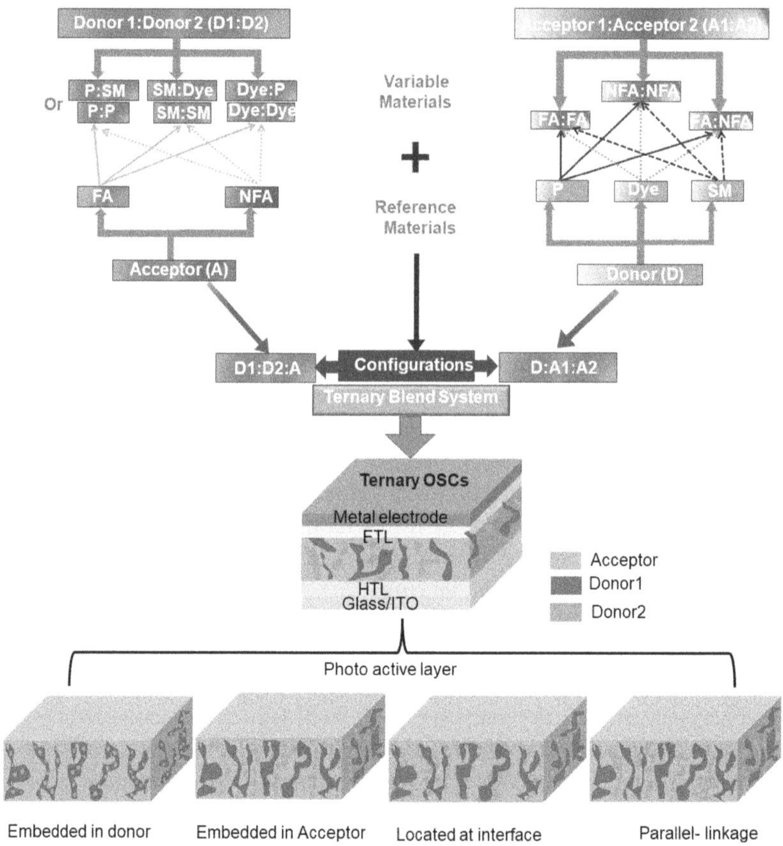

FIGURE 4.19 Ternary blend systems and possible active layer morphologies: possible combinations of the three elements: (Left) D1:D2:A and (Right) D:A1:A2 systems as an active layer. Here, the elements are limited to FA: fullerene acceptors, and NFA: Non-fullerene acceptors, P: polymers, SM: small molecules, and dyes.

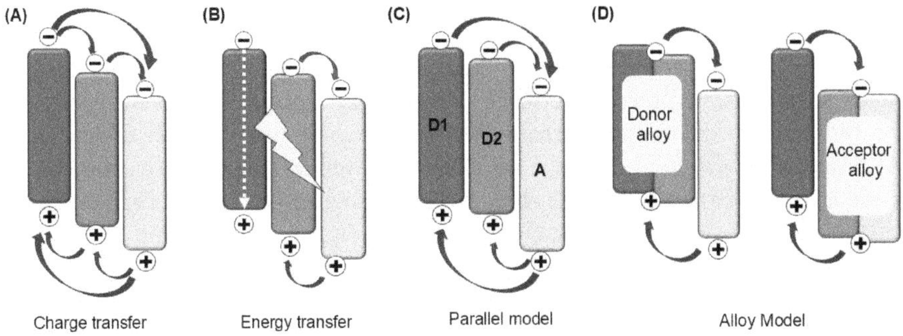

FIGURE 4.20 Visual illustration of energy and charge transfer pathways in the ternary blend organic solar cells.

acceptor, into the ternary active layer. Figure 4.20 presents the possible active layer morphologies according to the location of the third component. Here are the four possible configurations:

4.6.1.4.1 *Homogeneous Blend*

In this morphology, all three components are uniformly distributed throughout the active layer. The third component is mixed evenly with the donor and acceptor materials, creating a homogeneous blend. This configuration aims to promote efficient charge transfer and minimize exciton recombination by ensuring a balanced distribution of the materials.

4.6.1.4.2 *Donor-Enriched Blend*

In a donor-enriched blend, the third component is predominantly located within the donor-rich regions of the active layer. This morphology is achieved by incorporating the third component into the donor material or positioning it at the interface between the donor and acceptor. The donor-enriched blend enhances charge generation and transport by facilitating efficient exciton dissociation at the donor-acceptor interface.

4.6.1.4.3 *Acceptor-Enriched Blend*

Conversely, in an acceptor-enriched blend, the third component is primarily situated within the acceptor-rich regions of the active layer. This morphology can be achieved by integrating the third component into the acceptor material or positioning it at the interface between the donor and acceptor. The acceptor-enriched blend promotes efficient charge extraction and transport, leading to enhanced device performance.

4.6.1.4.4 *Interface-Localized Blend*

In an interface-localized blend, the third component is specifically localized at the interface between the donor and acceptor materials. This morphology aims to optimize charge separation and transport by facilitating efficient exciton dissociation and charge transfer at the interface. The interface-localized blend can improve device efficiency and stability by minimizing charge recombination losses.

4.6.2 Charge Transfer in Ternary OSCs

The charge transfer and transport in ternary solar cells are different from those of binary solar cells, which are governed by various mechanisms, such as the content of the third component, electronic energy levels, and bandgap of the three components, the third component location in the active layer as well as final microstructure of the active layers. In ternary OSCs, the charge transfer mechanism involves the generation, separation, and transport of excitons (bound electron-hole pairs) within the device's active layer, consisting of a blend of three components: two donors (D1 and D2) and an acceptor (A). When light is absorbed by the active layer, excitons are generated within the donors (D1 or D2). These excitons can then undergo dissociation at the interfaces between the donors and the acceptor (D1/A or D2/A), leading to the creation of free charge carriers—electrons and holes. The resulting holes typically migrate towards the anode, which is usually made of a material with high electron affinity, while electrons move towards the cathode, often composed of a material with low electron affinity. The charge separation and transport pathways within ternary OSCs are influenced by the energy level alignment of the three components. Efficient charge transfer requires suitable energy level offsets between the HOMO (highest occupied molecular orbital) of the donors and the LUMO (lowest unoccupied molecular orbital) of the acceptor. Additionally, energy level alignment between the two donors and the acceptor can facilitate charge transfer between donors, enhancing overall device performance. Figure 4.20 presents the charge and energy transfer pathways in the ternary OSCs consisting of the photo-active layer with different combinations as shown in Figure 4.19.

4.6.3 Performance and Efficiency of OPV Cells

Organic photovoltaic (OPV) cells have undergone significant advancements in efficiency over the years. Here's a chronological overview of some key milestones in OSC efficiency. The first OSC reported by Tang, C. W. (1986) consisting a two-layer (Planer junction) organic photovoltaic cell. Although the efficiency of this early organic solar cell was modest (around 1%), it laid the foundation for further research and development in the field [163]. In 1995,bulk heterojunction PSCs were first demonstrated by A. Heeger and R.H. Friend using polymer (MEH-PPV) and fullerene (C_{60}) blend with a PCE of 2.9% [168,169]. Up to 2005, the PCE of OSCs was about 5%, which was achieved by using the P3HT/PCBM system [170]. For the first time in 2013, the PCE of OSC crossed over 10% using a low band gap polymer PDTP-DFBT and $PC_{71}BM$ blend system [171].

Nonfullerene-based (NFA) OSCs have reached a milestone in achieving PCE of over 19.0% with single junction architecture [164,165,172] and raised up to 20.6% with tandem structure [173]. Luo et al. reported an auxiliary sequential deposition technique for fabricating the OSCs, by introducing an additional casting step of dithieno[3,2-b:2′,3′-d]thiophene between the sequential depositions of the donor (D18-Cl) and acceptor (L8-BO) layers, as shown in Figure 4.21a and b. This approach results in a significant performance enhancement, increasing from 15% to over 18% compared to blend casting and sequential deposition methods. Moreover, by incorporating a branched-chain-engineered acceptor named L8-BO-X, device performance to 19.04%, J_{SC} of 26.78, V_{OC} of 0.893 V due to enhanced intermolecular packing, better energy level alignment achieving top-tier values for binary OSCs, as depicted in

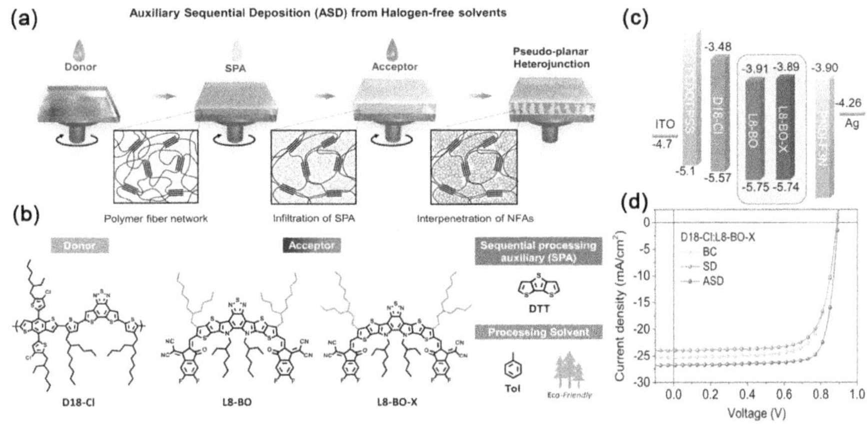

FIGURE 4.21 (a) Schematic illustrations of the device fabrication and possible nanoscale phase separation. (b) Chemical structure of the photoactive materials, D18-Cl, L8-BO, and L8-BO-X; (c) Energy level alignment of the materials and (d) *J–V* curves of D18-Cl:L8-BO-X from the BC, SD, and ASD methods. (Reproduced from Ref. [164] under the CC BY-4DEED license.)

Figure 4.21c and d [164]. Most recently, Liu et al. reported a single junction binary OSC utilizing NFA material called AQx-2F as the acceptor material has achieved a PCE of 19.7% [174]. This impressive performance was accompanied by a high V_{OC} of 0.937 V, a strong J_{SC} of 26.1 mA/cm², and a remarkable fill factor (FF) of 80.4%. These results position it as one of the highest-performing binary OSCs reported to date. NFAs offer a significant advantage over fullerene acceptors due to their enhanced light-absorbing capacity within the visible and near-infrared regions. This attribute contributes to the generation of heightened photocurrent in solar cells. The development of A-D-A type small molecule acceptors involves the synthesis and optimization of molecules with a specific architecture characterized by an electron-accepting core sandwiched between two electron-donating units. This design aims to enhance the efficiency of organic solar cells by facilitating efficient charge separation and transport. Zhan and colleagues have devised a range of exceptionally efficient A-D-A type small molecule acceptors [175]. Reversal review articles have been published on the development of these NFA-based A-D-A small molecules for OSC applications [176–179]. Additionally, ternary systems have demonstrated improved stability compared to their binary counterparts, making them more suitable for practical applications. Table 4.7 presents the performance of recently developed binary and ternary OSCs with PCE over 16%.

4.7 ISSUES ASSOCIATE WITH PSCs AND ORGANIC SOLAR CELLS

PSCS and OSCS have surged into the limelight within the photovoltaic realm due to their remarkable efficiencies, economical fabrication processes, and the tantalizing prospect of flexible and tandem device architectures. Nevertheless, a plethora of challenges accompany these promising attributes, mandating intensive research endeavors to surmount. Herein lies an extensive exposition of these challenges:

TABLE 4.7
Photovoltaic Performances of Selected Binary and Ternary OSCs

Type	Active Layer	V_{OC} (V)	J_{SC} (mA/cm^2)	FF (%)	PCE (%)	Working Mechanism	References
			Binary				
D:A	PM6:Y6	0.83	26.98	77.8	17.5	BHJ	[180]
D:A	PCE-10:IEICO-4F	0.840	26.65	78.1	17.48	GHJ	[181]
D:A	D18-Cl:BTP-4F-T3EH	0.873	26.84	77.85	18.25	BHJ	[182]
D:A	D18:L8-BO	0918	26.63	77.25	19.05	BHJ	[183]
D:A	PM6:BTP-eC9	0.861	27.88	80.39	19.31	BHJ	[165]
D:A	D18-Cl:L8-BO-X:PDI-F3N	0.893	26.78	79.6	19.04	PPHJ	[164]
D:A	AQx-2F	0.937	26.1	80.4	19.7	BHJ	[174]
			Ternary				
D1:D2:A	PBDT-ST:PNDT-ST:Y6-T	0.91	24.04	68.30	16.57	Alloy	[184]
D1:D2:A	PM6:J71:Y6	0.85	25.55	76.0	16.5	ET	[185]
D1:D2:A	PM6:PDHP-Th:Y6	0.85	26.60	71.7	16.8	CT	[186]
D1:D2:A	PM6:S3:Y6	0.86	25.86	79.17	17.53	Alloy	[187]
D:A1:A2	PM6:IDIC:Y6	0.868	25.39	74.92	16.50	Parallel	[188]
D:A1:A2	PM6:Y6:BTP-M	0.875	26.56	73.46	17.03	Alloy	[189]
D:A1:A2	PM6:Y18:PC71BM	0.84	26.3	77.4	17.11	—	[190]
D:A1:A2	PM6:BTP-eC9:BTP-F	0.858	26.99	79.7	18.45	Alloy	[191]
D:A1:A2	PM6:BTP-eC9:L8-BO-F	0.853	27.36	80.0	18.66	ET & Alloy	[192]
D:A1:A2	D18-Cl:G19:Y6	0.871	27.36	77.72	18.53	CT	[193]

Stability's sway: The Achilles' heel of perovskite materials lies in their susceptibility to degradation induced by the trifecta of moisture, oxygen, and thermal excursions, compounded by light exposure. This degradation culminates in diminished device efficiency and lifespan. Researchers are vigorously exploring diverse encapsulation methodologies, intricate interface engineering, and judicious material substitutions to fortify the stability of PSCs against such deleterious influences. One of the primary challenges with OSCs is improving their efficiency. While significant progress has been made in recent years, the PCE of OSCs still lags behind that of silicon and perovskite-based solar cells. Enhancing the efficiency involves optimizing the active layer materials, device architecture, and interface engineering.

Toxic tango: The presence of toxic materials, notably lead, within PSCs elicits grave environmental and health concerns. Consequently, concerted efforts are underway to forge lead-free or low-toxicity alternatives sans compromising device performance, thus fostering a more sustainable and ethically sound solar energy paradigm. Toxicity in OSCs primarily arises from the materials used in their fabrication. Many of these materials are considered relatively safe compared to inorganic alternatives, but some organic compounds used in OPVs may still have toxicity concerns. For instance, certain organic dyes or additives used for optimizing device performance may contain toxic elements or compounds. However, compared to traditional silicon-based solar cells that may contain heavy metals like lead and cadmium, organic solar cells generally have a lower toxicity footprint. The fabrication process of PSC and OSC typically involves the use of various solvents for dissolving perovskite and organic materials and depositing thin films. Some solvents commonly used in organic electronics, such as chlorinated solvents like DMF, DSMO, chloroform, or dichlorobenzene, are harmful to human health and the environment. Researchers are actively exploring alternative, less toxic solvents or developing solvent-free processing methods to mitigate these concerns.

Prodigious production predicaments: Despite the promise of low-cost fabrication via solution processing techniques, the Herculean task of scaling up production while preserving peak efficiency and reproducibility looms large. In this vein, advancements in manufacturing paradigms, such as the advent of roll-to-roll printing, are being fervently pursued to unlock the gates to large-scale commercialization.

Hysteresis hurdles in PSCs: The enigmatic phenomenon of hysteresis bedevils PSCs, manifesting as disparities in current-voltage characteristics and impeding the accurate determination of device efficiency. A comprehensive comprehension of hysteresis mechanisms coupled with adept mitigation strategies is imperative for attaining dependable performance metrics.

Lead's labyrinthine migration: The migration of lead ions or molecules within the perovskite layer or at interfaces poses a formidable threat, heralding device degradation and instability over time. Thus, strategies to either passivate or stabilize lead-containing interfaces constitute a paramount avenue of inquiry to circumvent this ominous specter.

Uniformity's unyielding quest: The quest for uniform perovskite film deposition and device fabrication stands as an imperative mandate for ensuring consistent device performance. Variances in film morphology, composition, and device architecture wield profound implications on device efficiency and reproducibility, necessitating meticulous attention.

Perovskite's ecological epilogue: The journey toward realizing the ecological promise of PSC and OSCs entails the formulation of robust strategies for recycling perovskite or organic materials and orchestrating their environmentally benign disposal at the culmination of the device's lifecycle. This conscientious approach serves as an indispensable linchpin in the overarching mission to minimize environmental impact and foster sustainability.

The rectification of these multifarious challenges constitutes an indispensable crucible for the widespread adoption of PSC and OSCs as a competitive and sustainable photovoltaic technology. Ongoing research and development endeavors, characterized by tenacity and innovation, are steadfastly directed toward unraveling these complexities and unleashing the full transformative potential of perovskite-based solar energy conversion.

4.8 CONCLUSIONS

PSCs and OSCs have witnessed a dramatic surge in efficiency, with recent advancements pushing their performance levels to approximately 26%, and ~20%, respectively, firmly establishing them as formidable contenders in the realm of solar energy alongside traditional wafer-based solar cells. Their inherent attributes, including cost-effectiveness, utilization of thin film technology, and reduced bulkiness, position them as the quintessential choice for the next generation of photovoltaic technology, promising to revolutionize the solar energy landscape.

The manufacturing methodologies employed in the production of PSCs, spanning liquid and vapor-assisted techniques, have undergone rigorous evaluation, each method unveiling unique strengths and limitations that directly impact the feasibility of large-scale production. Solution-based approaches, characterized by their swiftness and cost-effectiveness, offer enticing advantages yet grapple with issues related to the uniformity and stability of the resulting films. Conversely, vapor-based techniques, executed under meticulously controlled vacuum conditions, yield films with remarkable uniformity and high purity. However, these methods entail higher costs and slower deposition rates, posing challenges for large-scale production. Recognizing these considerations, there exists a notable emphasis on the development and implementation of roll-to-roll compatible methods, poised to streamline the fabrication process and facilitate the production of large, flexible PSCs and OSCs, thereby enhancing their commercial viability and widespread adoption. Moreover, ongoing research endeavors within the perovskite and OSCs domain are directed towards fortifying the long-term stability, an imperative step towards ensuring their durability and reliability in real-world applications. Concurrently, researchers are exploring the potential of novel, cost-effective materials to further optimize the performance and efficiency of PSCs/OSCs, with the aim of unlocking their full potential as a sustainable and competitive alternative in the solar energy sector.

As of my last update in January 2024, several startups and companies have emerged with the goal of commercializing PSC technology. These include Oxford Photovoltaics, Saule Technologies, Greatcell Solar, Hanwha Q CELLS, Enecoat Technologies, P3C (India), and Swift Solar, among others. These companies are actively working on scaling up production processes, improving stability, and integrating PSCs into commercial products and applications

NOTE

* Corresponding author: Neetesh Kumar, Department of Mechanical and Materials Engineering, NASA CRE2DO Center, Miami, Florida.

REFERENCES

1. J. Park et al., 'Controlled growth of perovskite layers with volatile alkylammonium chlorides', *Nature*, vol. 616, no. 7958, pp. 724–730, Apr. 2023, doi: 10.1038/s41586-023-05825-y
2. Z. Liang et al., 'Homogenizing out-of-plane cation composition in perovskite solar cells', *Nature*, vol. 624, no. 7992, pp. 557–563, Dec. 2023, doi: 10.1038/s41586-023-06784-0
3. J. Zhou et al., 'Highly efficient and stable perovskite solar cells via a multifunctional hole transporting material', *Joule*, vol. 8, pp. 1691–1706, Mar. 2024, doi: 10.1016/j.joule.2024.02.019
4. F. Wang, Y. Han, D. Duan, C. Ge, H. Hu, and G. Li, 'Recent progress of scalable perovskite solar cells and modules', *Energy Reviews*, vol. 1, no. 2, p. 100010, Dec. 2022, doi: 10.1016/j.enrev.2022.100010
5. J. Küffner, J. Hanisch, T. Wahl, J. Zillner, E. Ahlswede, and M. Powalla, 'One-step blade coating of inverted double-cation perovskite solar cells from a green precursor solvent', *ACS Applied Energy Materials*, vol. 4, no. 10, pp. 11700–11710, Oct. 2021, doi: 10.1021/acsaem.1c02425

6. W.-Q. Wu et al., 'Bilateral alkylamine for suppressing charge recombination and improving stability in blade-coated perovskite solar cells', *Science Advances*, vol. 5, no. 3, p. eaav8925, Mar. 2019, doi: 10.1126/sciadv.aav8925
7. S. Y. Abate et al., 'Room temperature slot-die coated perovskite layer modified with sulfonyl-γ-AApeptide for high performance perovskite solar devices', *Chemical Engineering Journal*, vol. 457, p. 141199, Feb. 2023, doi: 10.1016/j.cej.2022.141199
8. J. Li et al., 'Ink design enabling slot-die coated perovskite solar cells with >22% power conversion efficiency, micro-modules, and 1 year of outdoor performance evaluation', *Advanced Energy Materials*, vol. 13, no. 33, Sep. 2023, doi: 10.1002/aenm.202203898
9. B. Ding et al., 'Dopant-additive synergism enhances perovskite solar modules', *Nature*, vol. 628, pp. 299–305, Mar. 2024, doi: 10.1038/s41586-024-07228-z
10. R. Basu et al., 'Large-area organic photovoltaic modules with 14.5% certified world record efficiency', *Joule*, vol. 8, no. 4, pp. 970–978, Apr. 2024, doi: 10.1016/j.joule.2024.02.016
11. C. Han et al., 'Over 19% efficiency organic solar cells by regulating multidimensional intermolecular interactions', *Advanced Materials*, vol. 35, no. 10, Mar. 2023, doi: 10.1002/adma.202208986
12. C. Liu et al., 'Bimolecularly passivated interface enables efficient and stable inverted perovskite solar cells', *Science (1979)*, vol. 382, no. 6672, pp. 810–815, Nov. 2023, doi: 10.1126/science.adk1633
13. C. Zuo and L. Ding, 'Drop-casting to make efficient perovskite solar cells under high humidity', *Angewandte Chemie International Edition*, vol. 60, no. 20, pp. 11242–11246, May 2021, doi: 10.1002/anie.202101868
14. H. Xiao et al., 'Efficient inorganic perovskite solar cells made by drop-coating in ambient air', *Nano Energy*, vol. 106, p. 108061, Feb. 2023, doi: 10.1016/j.nanoen.2022.108061
15. C. Zuo et al., 'Crystallisation control of drop-cast quasi-2D/3D perovskite layers for efficient solar cells', *Communications Materials*, vol. 1, no. 1, p. 33, Jun. 2020, doi: 10.1038/s43246-020-0036-z
16. C. Zuo, A. D. Scully, and M. Gao, 'Drop-casting method to screen Ruddlesden–Popper perovskite formulations for use in solar cells', *ACS Applied Materials & Interfaces*, vol. 13, no. 47, pp. 56217–56225, Dec. 2021, doi: 10.1021/acsami.1c17475
17. Z. Bi et al., 'Fast preparation of uniform large grain size perovskite thin film in air condition via spray deposition method for high efficient planar solar cells', *Solar Energy Materials and Solar Cells*, vol. 162, pp. 13–20, Apr. 2017, doi: 10.1016/j.solmat.2016.12.032
18. K. M. Boopathi et al., 'Preparation of metal halide perovskite solar cells through a liquid droplet assisted method', *Journal of Materials Chemistry A*, vol. 3, no. 17, pp. 9257–9263, 2015, doi: 10.1039/C4TA06392A
19. C. F. J. Lau et al., '$CsPbIBr_2$ perovskite solar cell by spray-assisted deposition', *ACS Energy Letters*, vol. 1, no. 3, pp. 573–577, Sep. 2016, doi: 10.1021/acsenergylett.6b00341
20. F. Li et al., 'Microstructure modulation of the CH_3 NH_3 PbI_3 layer in perovskite solar cells by 2-propanol pre-wetting and annealing in a spray-assisted solution process', *Journal of Materials Chemistry A*, vol. 4, no. 29, pp. 11372–11380, 2016, doi: 10.1039/C6TA04600B
21. J. H. Heo, M. H. Lee, M. H. Jang, and S. H. Im, 'Highly efficient $CH_3NH_3PbI_{3-x}Cl_x$ mixed halide perovskite solar cells prepared by re-dissolution and crystal grain growth via spray coating', *Journal of Materials Chemistry A*, vol. 4, no. 45, pp. 17636–17642, 2016, doi: 10.1039/C6TA06718B
22. B. Tyagi et al., 'Development of high efficiency, spray-coated perovskite solar cells and modules using additive-engineered porous PbI_2 films', *Small Methods*, vol. 8, no. 2, Feb. 2024, doi: 10.1002/smtd.202300237

23. N. Kumar, H. B. Lee, S. Hwang, and J.-W. Kang, 'Large-area, green solvent spray deposited nickel oxide films for scalable fabrication of triple-cation perovskite solar cells', *Journal of Materials Chemistry A*, vol. 8, no. 6, pp. 3357–3368, 2020, doi: 10.1039/C9TA13528F
24. N. Kumar et al., 'Room-temperature spray deposition of large-area SnO_2 electron transport layer for high performance, stable $FAPbI_3$-based perovskite solar cells', *Small Methods*, vol. 6, no. 2, Feb. 2022, doi: 10.1002/smtd.202101127
25. Y.-S. Jung, K. Hwang, F. H. Scholes, S. E. Watkins, D.-Y. Kim, and D. Vak, 'Differentially pumped spray deposition as a rapid screening tool for organic and perovskite solar cells', *Scientific Reports*, vol. 6, no. 1, p. 20357, Feb. 2016, doi: 10.1038/srep20357
26. J. E. Bishop, T. J. Routledge, and D. G. Lidzey, 'Advances in spray-cast perovskite solar cells', *The Journal of Physical Chemistry Letters*, vol. 9, no. 8, pp. 1977–1984, Apr. 2018, doi: 10.1021/acs.jpclett.8b00311
27. P. S. Chandrasekhar, N. Kumar, S. K. Swami, V. Dutta, and V. K. Komarala, 'Fabrication of perovskite films using an electrostatic assisted spray technique: The effect of the electric field on morphology, crystallinity and solar cell performance', *Nanoscale*, vol. 8, no. 12, pp. 6792–6800, 2016, doi: 10.1039/C5NR08350H
28. J. E. Bishop, J. A. Smith, and D. G. Lidzey, 'Development of spray-coated perovskite solar cells', *ACS Applied Materials & Interfaces*, vol. 12, no. 43, pp. 48237–48245, Oct. 2020, doi: 10.1021/acsami.0c14540
29. A. Guo et al., 'Multi-channel pumped ultrasonic spray-coating for high-throughput and scalable mixed halide perovskite solar cells', *Advanced Materials Interfaces*, vol. 8, no. 5, Mar. 2021, doi: 10.1002/admi.202001509
30. J. H. Heo et al., 'Efficient and stable graded $CsPbI_{3-x}Br_x$ perovskite solar cells and submodules by orthogonal processable spray coating', *Joule*, vol. 5, no. 2, pp. 481–494, Feb. 2021, doi: 10.1016/j.joule.2020.12.010
31. H. Cai et al., 'High efficiency over 20% of perovskite solar cells by spray coating via a simple process', *ACS Applied Energy Materials*, vol. 3, no. 10, pp. 9696–9702, Oct. 2020, doi: 10.1021/acsaem.0c01129
32. Y.-S. Chou et al., 'Ultrasonic spray-coated mixed cation perovskite films and solar cells', *ACS Sustainable Chemistry & Engineering*, vol. 7, no. 16, pp. 14217–14224, Aug. 2019, doi: 10.1021/acssuschemeng.9b03058
33. E. J. Cassella et al., 'Ultrasonic spray deposition of a passivating agent for spray-coated, methylammonium-free perovskite solar cells', *Solar RRL*, vol. 8, no. 2, Jan. 2024, doi: 10.1002/solr.202300814
34. Z. Liu, G. Liu, C. Xu, and X. Xie, 'Scalable fabrication for efficient quasi two-dimensional perovskite solar cells via ultrasonic spray-coating method', *Organic Electronics*, vol. 102, p. 106440, Mar. 2022, doi: 10.1016/j.orgel.2022.106440
35. T. A. N. Peiris et al., 'Enhancement of the hole conducting effect of NiO by a N_2 blow drying method in printable perovskite solar cells with low-temperature carbon as the counter electrode', *Nanoscale*, vol. 9, no. 17, pp. 5475–5482, 2017, doi: 10.1039/C7NR00372B
36. M. Han, J. Byeon, J. Jang, C. Hur, G. Seo, and M. Choi, 'Hot gas-blowing assisted crystallinity management of bar-coated perovskite solar cells and modules', *Small Science*, vol. 3, no. 9, Sep. 2023, doi: 10.1002/smsc.202300069
37. E. Rezaee, D. Kutsarov, B. Li, J. Bi, and S. R. P. Silva, 'A route towards the fabrication of large-scale and high-quality perovskite films for optoelectronic devices', *Scientific Reports*, vol. 12, no. 1, p. 7411, May 2022, doi: 10.1038/s41598-022-10790-z
38. F. Jafarzadeh et al., 'All-blade-coated flexible perovskite solar cells & modules processed in air from a sustainable dimethyl sulfoxide (DMSO)-based solvent system', *Sustainable Energy & Fuels*, vol. 7, no. 9, pp. 2219–2228, 2023, doi: 10.1039/D2SE01678H
39. W.-Q. Wu et al., 'Molecular doping enabled scalable blading of efficient hole-transport-layer-free perovskite solar cells', *Nature Communications*, vol. 9, no. 1, p. 1625, Apr. 2018, doi: 10.1038/s41467-018-04028-8

40. Y. Lin et al., 'Manipulation of the crystallization of perovskite films induced by a rotating magnetic field during blade coating in air', *Journal of Materials Chemistry A*, vol. 6, no. 9, pp. 3986–3995, 2018, doi: 10.1039/C7TA09721B
41. Z. Yang, C. Chueh, F. Zuo, J. H. Kim, P. Liang, and A. K.-Y. Jen, 'High-performance fully printable perovskite solar cells via blade-coating technique under the ambient condition', *Advanced Energy Materials*, vol. 5, no. 13, Jul. 2015, doi: 10.1002/aenm.201500328
42. J. Li et al., 'Phase transition control for high-performance blade-coated perovskite solar cells', *Joule*, vol. 2, no. 7, pp. 1313–1330, Jul. 2018, doi: 10.1016/j.joule.2018.04.011
43. J. Zhang et al., 'Two-step sequential blade-coating of high-quality perovskite layers for efficient solar cells and modules', *Journal of Materials Chemistry A*, vol. 8, no. 17, pp. 8447–8454, 2020, doi: 10.1039/D0TA02043E
44. Z. Yang et al., 'Slot-die coating large-area formamidinium-cesium perovskite film for efficient and stable parallel solar module', *Science Advances*, vol. 7, no. 18, Apr. 2021, doi: 10.1126/sciadv.abg3749
45. S. S. Sangale, S. Kwon, P. Patil, H. Lee, and S. Na, 'Locally supersaturated inks for a slot-die process to enable highly efficient and robust perovskite solar cells', *Advanced Energy Materials*, vol. 13, no. 33, Sep. 2023, doi: 10.1002/aenm.202300537
46. J.-W. Lee, S.-I. Na, and S.-S. Kim, 'Efficient spin-coating-free planar heterojunction perovskite solar cells fabricated with successive brush-painting', *Journal of Power Sources*, vol. 339, pp. 33–40, Jan. 2017, doi: 10.1016/j.jpowsour.2016.11.028
47. S.-S. Kim, S.-I. Na, S.-J. Kang, and D.-Y. Kim, 'Annealing-free fabrication of P3HT:PCBM solar cells via simple brush painting', *Solar Energy Materials and Solar Cells*, vol. 94, no. 2, pp. 171–175, Feb. 2010, doi: 10.1016/j.solmat.2009.08.017
48. S.-S. Kim, S.-I. Na, J. Jo, G. Tae, and D.-Y. Kim, 'Efficient polymer solar cells fabricated by simple brush painting', *Advanced Materials*, vol. 19, no. 24, pp. 4410–4415, Dec. 2007, doi: 10.1002/adma.200702040
49. J. Huang et al., 'Direct conversion of $CH_3NH_3PbI_3$ from electrodeposited PbO for highly efficient planar perovskite solar cells', *Scientific Reports*, vol. 5, no. 1, p. 15889, Oct. 2015, doi: 10.1038/srep15889
50. J. C. Hill, J. A. Koza, and J. A. Switzer, 'Electrodeposition of epitaxial lead iodide and conversion to textured methylammonium lead iodide perovskite', *ACS Applied Materials & Interfaces*, vol. 7, no. 47, pp. 26012–26016, Dec. 2015, doi: 10.1021/acsami.5b07222
51. I. Kosta, H. Grande, and R. Tena-Zaera, 'Dimethylformamide-free processing of halide perovskite solar cells from electrodeposited PbI_2 precursor films', *Electrochim Acta*, vol. 246, pp. 1193–1199, Aug. 2017, doi: 10.1016/j.electacta.2017.06.104
52. W. Li, J. Yang, Q. Jiang, R. Li, and L. Zhao, 'Electrochemical deposition of PbI_2 for perovskite solar cells', *Solar Energy*, vol. 159, pp. 300–305, Jan. 2018, doi: 10.1016/j.solener.2017.10.077
53. M. Al-Katrib, L. Perrin, L. Flandin, and E. Planes, 'Electrodeposition in perovskite solar cells: A critical review, new insights, and promising paths to future industrial applications', *Advanced Materials Technologies*, vol. 8, no. 23, Dec. 2023, doi: 10.1002/admt.202300964
54. F. Schackmar et al., 'Perovskite solar cells with all-inkjet-printed absorber and charge transport layers', *Advanced Materials Technologies*, vol. 6, no. 2, Feb. 2021, doi: 10.1002/admt.202000271
55. T. Abzieher et al., 'Electron-beam-evaporated nickel oxide hole transport layers for perovskite-based photovoltaics', *Advanced Energy Materials*, vol. 9, no. 12, Mar. 2019, doi: 10.1002/aenm.201802995
56. P. Li et al., 'Inkjet manipulated homogeneous large size perovskite grains for efficient and large-area perovskite solar cells', *Nano Energy*, vol. 46, pp. 203–211, Apr. 2018, doi: 10.1016/j.nanoen.2018.01.049

57. Z. Li et al., 'Ink engineering of inkjet printing perovskite', *ACS Applied Materials & Interfaces*, vol. 12, no. 35, pp. 39082–39091, Sep. 2020, doi: 10.1021/acsami.0c09485
58. V. V. Satale et al., 'Spray-assisted deposition of a SnO_2 electron transport bilayer for efficient inkjet-printed perovskite solar cells', *Inorganic Chemistry Frontiers*, vol. 10, no. 12, pp. 3558–3567, 2023, doi: 10.1039/D3QI00599B
59. M. Liu, M. B. Johnston, and H. J. Snaith, 'Efficient planar heterojunction perovskite solar cells by vapour deposition', *Nature*, vol. 501, no. 7467, pp. 395–398, Sep. 2013, doi: 10.1038/nature12509
60. K. B. Lohmann et al., 'Control over crystal size in vapor deposited metal-halide perovskite films', *ACS Energy Letters*, vol. 5, no. 3, pp. 710–717, Mar. 2020, doi: 10.1021/acsenergylett.0c00183
61. L. Yang et al., 'High-quality conformal homogeneous all-vacuum deposited $CsPbCl_3$ thin films and their UV photodiode applications', *ACS Applied Materials & Interfaces*, vol. 11, no. 50, pp. 47054–47062, Dec. 2019, doi: 10.1021/acsami.9b16264
62. D. Lin et al., 'Stable and scalable 3D-2D planar heterojunction perovskite solar cells via vapor deposition', *Nano Energy*, vol. 59, pp. 619–625, May 2019, doi: 10.1016/j.nanoen.2019.03.014
63. B. Gao et al., 'Organic-inorganic perovskite films and efficient planar heterojunction solar cells by magnetron sputtering', *Advanced Science*, vol. 8, no. 22, Nov. 2021, doi: 10.1002/advs.202102081
64. B. Gao et al., 'Doping mechanism of perovskite films with $PbCl_2$ prepared by magnetron sputtering for enhanced efficiency of solar cells', *ACS Applied Materials & Interfaces*, vol. 14, no. 35, pp. 40062–40071, Sep. 2022, doi: 10.1021/acsami.2c11929
65. F. Xu et al., 'Large-area $CsPbBr_3$ perovskite films grown with effective one-step RF-magnetron sputtering', *Journal of Applied Physics*, vol. 129, no. 24, Jun. 2021, doi: 10.1063/5.0050810
66. W. A. Dunlap-Shohl, E. T. Barraza, A. Barrette, K. Gundogdu, A. D. Stiff-Roberts, and D. B. Mitzi, '$MAPbI_3$ solar cells with absorber deposited by resonant infrared matrix-assisted pulsed laser evaporation', *ACS Energy Letters*, vol. 3, no. 2, pp. 270–275, Feb. 2018, doi: 10.1021/acsenergylett.7b01144
67. H. Wang et al., 'Pulsed laser deposition of $CsPbBr_3$ films for application in perovskite solar cells', *ACS Applied Materials & Interfaces*, vol. 2, no. 3, pp. 2305–2312, Mar. 2019, doi: 10.1021/acsaem.9b00130
68. N. Rodkey et al., 'Pulsed laser deposition of Cs_2 $AgBiBr_6$: From mechanochemically synthesized powders to dry, single-step deposition', *Chemistry of Materials*, vol. 33, no. 18, pp. 7417–7422, Sep. 2021, doi: 10.1021/acs.chemmater.1c02054
69. Y. Jiang et al., 'Negligible-Pb-waste and upscalable perovskite deposition technology for high-operational-stability perovskite solar modules', *Advanced Energy Materials*, vol. 9, no. 13, Apr. 2019, doi: 10.1002/aenm.201803047
70. J. Liu et al., 'Crystalline quality control in sequential vapor deposited perovskite film toward high efficiency and large scale solar cells', *Solar Energy Materials and Solar Cells*, vol. 233, p. 111382, Dec. 2021, doi: 10.1016/j.solmat.2021.111382
71. L. Qiu et al., 'Rapid hybrid chemical vapor deposition for efficient and hysteresis-free perovskite solar modules with an operation lifetime exceeding 800 hours', *Journal of Materials Chemistry A*, vol. 8, no. 44, pp. 23404–23412, 2020, doi: 10.1039/D0TA09007G
72. M. M. Tavakoli et al., 'Fabrication of efficient planar perovskite solar cells using a one-step chemical vapor deposition method', *Scientific Reports*, vol. 5, no. 1, p. 14083, Sep. 2015, doi: 10.1038/srep14083
73. M. Li et al., 'Highly efficient 2D/3D hybrid perovskite solar cells via low-pressure vapor-assisted solution process', *Advanced Materials*, vol. 30, no. 30, Jul. 2018, doi: 10.1002/adma.201801401

74. Y. Jiang, S. He, L. Qiu, Y. Zhao, and Y. Qi, 'Perovskite solar cells by vapor deposition based and assisted methods', *Applied Physics Reviews*, vol. 9, no. 2, Jun. 2022, doi: 10.1063/5.0085221
75. Q. Chen et al., 'Planar heterojunction perovskite solar cells via vapor-assisted solution process', *Journal of the American Chemical Society*, vol. 136, no. 2, pp. 622–625, Jan. 2014, doi: 10.1021/ja411509g
76. X. Zhu et al., 'Vapor-fumigation for record efficiency two-dimensional perovskite solar cells with superior stability', *Energy & Environmental Science*, vol. 11, no. 12, pp. 3349–3357, 2018, doi: 10.1039/C8EE02284D
77. X. Zheng et al., 'Managing grains and interfaces via ligand anchoring enables 22.3%-efficiency inverted perovskite solar cells', *Nature Energy*, vol. 5, no. 2, pp. 131–140, Jan. 2020, doi: 10.1038/s41560-019-0538-4
78. H. Lu et al., 'Vapor-assisted deposition of highly efficient, stable black-phase $FAPbI_3$ perovskite solar cells', *Science (1979)*, vol. 370, no. 6512, Oct. 2020, doi: 10.1126/science.abb8985
79. Y. Xu et al., 'Synchronous modulation of defects and buried interfaces for highly efficient inverted perovskite solar cells', *Advanced Energy Materials*, vol. 13, no. 3, Jan. 2023, doi: 10.1002/aenm.202203505
80. W. Qiu et al., 'Low-temperature robust $MAPbI_3$ perovskite solar cells with power conversion efficiency exceeding 22.4%', *Chemical Engineering Journal*, vol. 468, p. 143656, Jul. 2023, doi: 10.1016/j.cej.2023.143656
81. T. Das, G. Di Liberto, and G. Pacchioni, 'Density functional theory estimate of halide perovskite band gap: When spin orbit coupling helps', *The Journal of Physical Chemistry* C, vol. 126, no. 4, pp. 2184–2198, Feb. 2022, doi: 10.1021/acs.jpcc.1c09594
82. P. Ahlawat et al., 'A combined molecular dynamics and experimental study of two-step process enabling low-temperature formation of phase-pure α-$FAPbI_3$', *Science Advances*, vol. 7, no. 17, Apr. 2021, doi: 10.1126/sciadv.abe3326
83. M. R. A. Elsayed, A. M. Elseman, A. A. Abdelmageed, H. M. Hashem, and A. Hassen, 'Synthesis and numerical simulation of formamidinium-based perovskite solar cells: A predictable device performance at NIS-Egypt', *Scientific Reports*, vol. 13, no. 1, p. 10115, Jun. 2023, doi: 10.1038/s41598-023-37018-y
84. L. Zhang et al., 'Advances in the application of perovskite materials', *Nanomicro Letters*, vol. 15, no. 1, p. 177, Dec. 2023, doi: 10.1007/s40820-023-01140-3
85. K. Galkowski et al., 'Determination of the exciton binding energy and effective masses for methylammonium and formamidinium lead tri-halide perovskite semiconductors', *Energy & Environmental Science*, vol. 9, no. 3, pp. 962–970, 2016, doi: 10.1039/C5EE03435C
86. Q. Jiang et al., 'Enhanced electron extraction using SnO_2 for high-efficiency planar-structure $HC(NH_2)_2PbI_3$-based perovskite solar cells', *Nature Energy*, vol. 2, no. 1, p. 16177, Nov. 2016, doi: 10.1038/nenergy.2016.177
87. X. Cui, J. Jin, Q. Tai, and F. Yan, 'Recent progress on the phase stabilization of $FAPbI_3$ for high-performance perovskite solar cells', *Solar RRL*, vol. 6, no. 10, Oct. 2022, doi: 10.1002/solr.202200497
88. M. RaeisianAsl, S. F. K. S. Panahi, M. Jamaati, and S. S. Tafreshi, 'A review on theoretical studies of structural and optoelectronic properties of FA-based perovskite materials with a focus on $FAPbI_3$', *International Journal of Energy Research*, vol. 46, no. 10, pp. 13117–13151, Aug. 2022, doi: 10.1002/er.8008
89. Q. Han et al., 'Single crystal formamidinium lead iodide ($FAPbI_3$): Insight into the structural, optical, and electrical properties', *Advanced Materials*, vol. 28, no. 11, pp. 2253–2258, Mar. 2016, doi: 10.1002/adma.201505002

90. B. Kim, J. Kim, and N. Park, 'First-principles identification of the charge-shifting mechanism and ferroelectricity in hybrid halide perovskites', *Scientific Reports*, vol. 10, no. 1, p. 19635, Nov. 2020, doi: 10.1038/s41598-020-76742-7
91. N. J. Jeon et al., 'Compositional engineering of perovskite materials for high-performance solar cells', *Nature*, vol. 517, no. 7535, pp. 476–480, Jan. 2015, doi: 10.1038/nature14133
92. M. Kim et al., 'Methylammonium chloride induces intermediate phase stabilization for efficient perovskite solar cells', *Joule*, vol. 3, no. 9, pp. 2179–2192, Sep. 2019, doi: 10.1016/j.joule.2019.06.014
93. M. Saliba et al., 'Cesium-containing triple cation perovskite solar cells: Improved stability, reproducibility and high efficiency', *Energy & Environmental Science*, vol. 9, no. 6, pp. 1989–1997, 2016, doi: 10.1039/C5EE03874J
94. I. Susic, L. Gil-Escrig, F. Palazon, M. Sessolo, and H. J. Bolink, 'Quadruple-cation wide-bandgap perovskite solar cells with enhanced thermal stability enabled by vacuum deposition', *ACS Energy Letters*, vol. 7, no. 4, pp. 1355–1363, Apr. 2022, doi: 10.1021/acsenergylett.2c00304
95. Z. Huang et al., 'Anion–π interactions suppress phase impurities in $FAPbI_3$ solar cells', *Nature*, vol. 623, no. 7987, pp. 531–537, Nov. 2023, doi: 10.1038/s41586-023-06637-w
96. A. Kojima, K. Teshima, Y. Shirai, and T. Miyasaka, 'Organometal halide perovskites as visible-light sensitizers for photovoltaic cells', *Journal of the American Chemical Society*, vol. 131, no. 17, pp. 6050–6051, May 2009, doi: 10.1021/ja809598r
97. J.-H. Im, C.-R. Lee, J.-W. Lee, S.-W. Park, and N.-G. Park, '6.5% efficient perovskite quantum-dot-sensitized solar cell', *Nanoscale*, vol. 3, no. 10, p. 4088, 2011, doi: 10.1039/c1nr10867k
98. H.-S. Kim et al., 'Lead iodide perovskite sensitized all-solid-state submicron thin film mesoscopic solar cell with efficiency exceeding 9%', *Scientific Reports*, vol. 2, no. 1, p. 591, Aug. 2012, doi: 10.1038/srep00591
99. M. M. Lee, J. Teuscher, T. Miyasaka, T. N. Murakami, and H. J. Snaith, 'Efficient hybrid solar cells based on meso-superstructured organometal halide perovskites', *Science (1979)*, vol. 338, no. 6107, pp. 643–647, Nov. 2012, doi: 10.1126/science.1228604
100. J. M. Ball, M. M. Lee, A. Hey, and H. J. Snaith, 'Low-temperature processed meso-superstructured to thin-film perovskite solar cells', *Energy & Environmental Science*, vol. 6, no. 6, p. 1739, 2013, doi: 10.1039/c3ee40810h
101. H. Zhou et al., 'Interface engineering of highly efficient perovskite solar cells', *Science (1979)*, vol. 345, no. 6196, pp. 542–546, Aug. 2014, doi: 10.1126/science.1254050
102. N. Ahn, D.-Y. Son, I.-H. Jang, S. M. Kang, M. Choi, and N.-G. Park, 'Highly reproducible perovskite solar cells with average efficiency of 18.3% and best efficiency of 19.7% fabricated via Lewis base adduct of lead(II) iodide', *Journal of the American Chemical Society*, vol. 137, no. 27, pp. 8696–8699, Jul. 2015, doi: 10.1021/jacs.5b04930
103. D.-Y. Son et al., 'Self-formed grain boundary healing layer for highly efficient $CH_3NH_3PbI_3$ perovskite solar cells', *Nature Energy*, vol. 1, no. 7, p. 16081, Jun. 2016, doi: 10.1038/nenergy.2016.81
104. S. S. Shin et al., 'Colloidally prepared La-doped $BaSnO_3$ electrodes for efficient, photostable perovskite solar cells', *Science (1979)*, vol. 356, no. 6334, pp. 167–171, Apr. 2017, doi: 10.1126/science.aam6620
105. A. Gerwien, M. Schildhauer, S. Thumser, P. Mayer, and H. Dube, 'Direct evidence for hula twist and single-bond rotation photoproducts', *Nature Communications*, vol. 9, no. 1, p. 2510, Jun. 2018, doi: 10.1038/s41467-018-04928-9
106. Y. Wang et al., 'Dopant-free small-molecule hole-transporting material for inverted perovskite solar cells with efficiency exceeding 21%', *Advanced Materials*, vol. 31, no. 35, Aug. 2019, doi: 10.1002/adma.201902781

107. K. Wang et al., 'Isothermally crystallized perovskites at room-temperature', *Energy & Environmental Science*, vol. 13, no. 10, pp. 3412–3422, 2020, doi: 10.1039/D0EE01967D
108. T. M. Koh et al., 'Formamidinium-containing metal-halide: An alternative material for near-IR absorption perovskite solar cells', *The Journal of Physical Chemistry* C, vol. 118, no. 30, pp. 16458–16462, Jul. 2014, doi: 10.1021/jp411112k
109. J. Lee, D. Seol, A. Cho, and N. Park, 'High-efficiency perovskite solar cells based on the black polymorph of $HC(NH_2)_2$ PbI_3', *Advanced Materials*, vol. 26, no. 29, pp. 4991–4998, Aug. 2014, doi: 10.1002/adma.201401137
110. W. S. Yang et al., 'High-performance photovoltaic perovskite layers fabricated through intramolecular exchange', *Science (1979)*, vol. 348, no. 6240, pp. 1234–1237, Jun. 2015, doi: 10.1126/science.aaa9272
111. D. Bi et al., 'Polymer-templated nucleation and crystal growth of perovskite films for solar cells with efficiency greater than 21%', *Nature Energy*, vol. 1, no. 10, p. 16142, Sep. 2016, doi: 10.1038/nenergy.2016.142
112. W. S. Yang et al., 'Iodide management in formamidinium-lead-halide–based perovskite layers for efficient solar cells', *Science (1979)*, vol. 356, no. 6345, pp. 1376–1379, Jun. 2017, doi: 10.1126/science.aan2301
113. S.-H. Turren-Cruz, A. Hagfeldt, and M. Saliba, 'Methylammonium-free, high-performance, and stable perovskite solar cells on a planar architecture', *Science (1979)*, vol. 362, no. 6413, pp. 449–453, Oct. 2018, doi: 10.1126/science.aat3583
114. H. Min et al., 'Efficient, stable solar cells by using inherent bandgap of α-phase formamidinium lead iodide', *Science (1979)*, vol. 366, no. 6466, pp. 749–753, Nov. 2019, doi: 10.1126/science.aay7044
115. N. Li et al., 'Liquid medium annealing for fabricating durable perovskite solar cells with improved reproducibility', *Science (1979)*, vol. 373, no. 6554, pp. 561–567, Jul. 2021, doi: 10.1126/science.abh3884
116. J. Jeong et al., 'Pseudo-halide anion engineering for α-$FAPbI_3$ perovskite solar cells', *Nature*, vol. 592, no. 7854, pp. 381–385, Apr. 2021, doi: 10.1038/s41586-021-03406-5
117. J. J. Yoo et al., 'Efficient perovskite solar cells via improved carrier management', *Nature*, vol. 590, no. 7847, pp. 587–593, Feb. 2021, doi: 10.1038/s41586-021-03285-w
118. M. J. Jeong et al., 'Boosting radiation of stacked halide layer for perovskite solar cells with efficiency over 25%', *Joule*, vol. 7, no. 1, pp. 112–127, Jan. 2023, doi: 10.1016/j.joule.2022.10.015
119. Z. Li et al., 'Stabilized hole-selective layer for high-performance inverted p-i-n perovskite solar cells', *Science (1979)*, vol. 382, no. 6668, pp. 284–289, Oct. 2023, doi: 10.1126/science.ade9637
120. H. Min et al., 'Perovskite solar cells with atomically coherent interlayers on SnO_2 electrodes', *Nature*, vol. 598, no. 7881, pp. 444–450, Oct. 2021, doi: 10.1038/s41586-021-03964-8
121. M. Kim et al., 'Conformal quantum dot–SnO_2 layers as electron transporters for efficient perovskite solar cells', *Science (1979)*, vol. 375, no. 6578, pp. 302–306, Jan. 2022, doi: 10.1126/science.abh1885
122. Y. Zhao et al., 'Inactive $(PbI_2)_2$ RbCl stabilizes perovskite films for efficient solar cells', *Science (1979)*, vol. 377, no. 6605, pp. 531–534, Jul. 2022, doi: 10.1126/science.abp8873
123. Q. Jiang et al., 'Surface reaction for efficient and stable inverted perovskite solar cells', *Nature*, vol. 611, no. 7935, pp. 278–283, Nov. 2022, doi: 10.1038/s41586-022-05268-x
124. Z. Li et al., 'Organometallic-functionalized interfaces for highly efficient inverted perovskite solar cells', *Science (1979)*, vol. 376, no. 6591, pp. 416–420, Apr. 2022, doi: 10.1126/science.abm8566

125. P. Shi et al., 'Oriented nucleation in formamidinium perovskite for photovoltaics', *Nature*, vol. 620, no. 7973, pp. 323–327, Aug. 2023, doi: 10.1038/s41586-023-06208-z
126. C. Ma et al., 'Photovoltaically top-performing perovskite crystal facets', *Joule*, vol. 6, no. 11, pp. 2626–2643, Nov. 2022, doi: 10.1016/j.joule.2022.09.012
127. T. Zhang et al., 'Ion-modulated radical doping of spiro-OMeTAD for more efficient and stable perovskite solar cells', *Science (1979)*, vol. 377, no. 6605, pp. 495–501, Jul. 2022, doi: 10.1126/science.abo2757
128. W. Yang et al., 'Visualizing interfacial energy offset and defects in efficient 2D/3D heterojunction perovskite solar cells and modules', *Advanced Materials*, vol. 35, no. 35, Sep. 2023, doi: 10.1002/adma.202302071
129. S. Zhang et al., 'Minimizing buried interfacial defects for efficient inverted perovskite solar cells', *Science (1979)*, vol. 380, no. 6643, pp. 404–409, Apr. 2023, doi: 10.1126/science.adg3755
130. Q. Tan et al., 'Inverted perovskite solar cells using dimethylacridine-based dopants', *Nature*, vol. 620, no. 7974, pp. 545–551, Aug. 2023, doi: 10.1038/s41586-023-06207-0
131. B. Jiao et al., 'Realizing stable perovskite solar cells with efficiency exceeding 25.6% through crystallization kinetics and spatial orientation regulation', *Advanced Materials*, vol. 36, Mar. 2024, doi: 10.1002/adma.202313673
132. N. Pellet et al., 'Mixed-organic-cation perovskite photovoltaics for enhanced solar-light harvesting', *Angewandte Chemie International Edition*, vol. 53, no. 12, pp. 3151–3157, Mar. 2014, doi: 10.1002/anie.201309361
133. M. Saliba et al., 'Incorporation of rubidium cations into perovskite solar cells improves photovoltaic performance', *Science (1979)*, vol. 354, no. 6309, pp. 206–209, Oct. 2016, doi: 10.1126/science.aah5557
134. A. Mahapatra, D. Prochowicz, M. M. Tavakoli, S. Trivedi, P. Kumar, and P. Yadav, 'A review of aspects of additive engineering in perovskite solar cells', *Journal of Materials Chemistry A*, vol. 8, no. 1, pp. 27–54, 2020, doi: 10.1039/C9TA07657C
135. C. Pereyra, H. Xie, and M. Lira-Cantu, 'Additive engineering for stable halide perovskite solar cells', *Journal of Energy Chemistry*, vol. 60, pp. 599–634, Sep. 2021, doi: 10.1016/j.jechem.2021.01.037
136. H. Wang et al., 'Additive engineering for mixed lead–tin narrow-band-gap perovskite solar cells: Recent advances and perspectives', *Energy & Fuels*, vol. 37, no. 9, pp. 6401–6423, May 2023, doi: 10.1021/acs.energyfuels.3c00462
137. Z. Li, M. Yang, J.-S. Park, S.-H. Wei, J. J. Berry, and K. Zhu, 'Stabilizing perovskite structures by tuning tolerance factor: Formation of formamidinium and cesium lead iodide solid-state alloys', *Chemistry of Materials*, vol. 28, no. 1, pp. 284–292, Jan. 2016, doi: 10.1021/acs.chemmater.5b04107
138. N.-K. Cho, H.-J. Na, J. Yoo, and Y. S. Kim, 'Long-term stability in γ-$CsPbI_3$ perovskite via an ultraviolet-curable polymer network', *Communications Materials*, vol. 2, no. 1, p. 30, Mar. 2021, doi: 10.1038/s43246-021-00134-1
139. J. Wang et al., '21.15%-Efficiency and stable γ-$CsPbI_3$ perovskite solar cells enabled by an acyloin ligand', *Advanced Materials*, vol. 35, no. 12, Mar. 2023, doi: 10.1002/adma.202210223
140. H. Zhu et al., 'Long-term operating stability in perovskite photovoltaics', *Nature Reviews Materials*, vol. 8, no. 9, pp. 569–586, Aug. 2023, doi: 10.1038/s41578-023-00582-w
141. G. E. Eperon et al., 'The role of dimethylammonium in bandgap modulation for stable halide perovskites', *ACS Energy Letters*, vol. 5, no. 6, pp. 1856–1864, Jun. 2020, doi: 10.1021/acsenergylett.0c00872
142. X. Dong, D. Chen, J. Zhou, Y.-Z. Zheng, and X. Tao, 'High crystallization of a multiple cation perovskite absorber for low-temperature stable ZnO solar cells with high-efficiency of over 20%', *Nanoscale*, vol. 10, no. 15, pp. 7218–7227, 2018, doi: 10.1039/C8NR00152A

143. Z. Chen, X. Zheng, F. Yao, J. Ma, C. Tao, and G. Fang, 'Methylammonium, formamidinium and ethylenediamine mixed triple-cation perovskite solar cells with high efficiency and remarkable stability', *Journal of Materials Chemistry A*, vol. 6, no. 36, pp. 17625–17632, 2018, doi: 10.1039/C8TA05756G
144. S. Yu, H. Liu, S. Wang, H. Zhu, X. Dong, and X. Li, 'Hydrazinium cation mixed $FAPbI_3$-based perovskite with 1D/3D hybrid dimension structure for efficient and stable solar cells', *Chemical Engineering Journal*, vol. 403, p. 125724, Jan. 2021, doi: 10.1016/j.cej.2020.125724
145. J. Cao et al., 'Alkali-cation-enhanced benzylammonium passivation for efficient and stable perovskite solar cells fabricated through sequential deposition', *Journal of Materials Chemistry A*, vol. 8, no. 37, pp. 19357–19366, 2020, doi: 10.1039/D0TA04680A
146. T. Bu et al., 'Sub-sized monovalent alkaline cations enhanced electrical stability for over 17% hysteresis-free planar perovskite solar mini-module', *Electrochim Acta*, vol. 306, pp. 635–642, May 2019, doi: 10.1016/j.electacta.2019.03.176
147. A. D. Jodlowski et al., 'Large guanidinium cation mixed with methylammonium in lead iodide perovskites for 19% efficient solar cells', *Nature Energy*, vol. 2, no. 12, pp. 972–979, Dec. 2017, doi: 10.1038/s41560-017-0054-3
148. H. B. Lee, N. Kumar, B. Tyagi, S. He, R. Sahani, and J.-W. Kang, 'Bulky organic cations engineered lead-halide perovskites: A review on dimensionality and optoelectronic applications', *Materials Today Energy*, vol. 21, p. 100759, Sep. 2021, doi: 10.1016/j.mtener.2021.100759
149. T. He et al., 'Reduced-dimensional perovskite photovoltaics with homogeneous energy landscape', *Nature Communications*, vol. 11, no. 1, p. 1672, Apr. 2020, doi: 10.1038/s41467-020-15451-1
150. O. Bubnova, 'Low-dimensional perovskites', *Nature Nanotechnology*, vol. 13, no. 7, pp. 531–531, Jul. 2018, doi: 10.1038/s41565-018-0214-z
151. I. Metcalf et al., 'Synergy of 3D and 2D perovskites for durable, efficient solar cells and beyond', *Chemical Reviews*, vol. 123, no. 15, pp. 9565–9652, Aug. 2023, doi: 10.1021/acs.chemrev.3c00214
152. M. A. Uddin et al., 'Iodide manipulation using zinc additives for efficient perovskite solar minimodules', *Nature Communications*, vol. 15, no. 1, p. 1355, Feb. 2024, doi: 10.1038/s41467-024-45649-6
153. H. C. Weerasinghe et al., 'The first demonstration of entirely roll-to-roll fabricated perovskite solar cell modules under ambient room conditions', *Nature Communications*, vol. 15, no. 1, p. 1656, Mar. 2024, doi: 10.1038/s41467-024-46016-1
154. Z. Liu et al., 'A holistic approach to interface stabilization for efficient perovskite solar modules with over 2,000-hour operational stability', *Nature Energy*, vol. 5, no. 8, pp. 596–604, Jul. 2020, doi: 10.1038/s41560-020-0653-2
155. K. S. Keremane et al., 'Improving the performance of carbon-based perovskite solar modules (70 cm^2) by incorporating cesium halide in mesoporous TiO_2', *ACS Applied Energy Materials*, vol. 4, no. 1, pp. 249–258, Jan. 2021, doi: 10.1021/acsaem.0c02213
156. C. Liu et al., 'Tuning structural isomers of phenylenediammonium to afford efficient and stable perovskite solar cells and modules', *Nature Communications*, vol. 12, no. 1, p. 6394, Nov. 2021, doi: 10.1038/s41467-021-26754-2
157. Y. Deng, S. Xu, S. Chen, X. Xiao, J. Zhao, and J. Huang, 'Defect compensation in formamidinium–caesium perovskites for highly efficient solar mini-modules with improved photostability', *Nature Energy*, vol. 6, no. 6, pp. 633–641, May 2021, doi: 10.1038/s41560-021-00831-8
158. T. Bu et al., 'Lead halide–templated crystallization of methylamine-free perovskite for efficient photovoltaic modules', *Science (1979)*, vol. 372, no. 6548, pp. 1327–1332, Jun. 2021, doi: 10.1126/science.abh1035

159. S. Chen, X. Dai, S. Xu, H. Jiao, L. Zhao, and J. Huang, 'Stabilizing perovskite-substrate interfaces for high-performance perovskite modules', *Science (1979)*, vol. 373, no. 6557, pp. 902–907, Aug. 2021, doi: 10.1126/science.abi6323
160. J. Xia et al., 'Tailoring electric dipole of hole-transporting material p-dopants for perovskite solar cells', *Joule*, vol. 6, no. 7, pp. 1689–1709, Jul. 2022, doi: 10.1016/j.joule.2022.05.012
161. S. You et al., 'Radical polymeric p-doping and grain modulation for stable, efficient perovskite solar modules', *Science (1979)*, vol. 379, no. 6629, pp. 288–294, Jan. 2023, doi: 10.1126/science.add8786
162. L. Luo et al., 'Stabilization of 3D/2D perovskite heterostructures via inhibition of ion diffusion by cross-linked polymers for solar cells with improved performance', *Nature Energy*, vol. 8, no. 3, pp. 294–303, Feb. 2023, doi: 10.1038/s41560-023-01205-y
163. C. W. Tang, 'Two-layer organic photovoltaic cell', *Applied Physics Letters*, vol. 48, no. 2, pp. 183–185, Jan. 1986, doi: 10.1063/1.96937
164. S. Luo et al., 'Auxiliary sequential deposition enables 19%-efficiency organic solar cells processed from halogen-free solvents', *Nature Communications*, vol. 14, no. 1, p. 6964, Oct. 2023, doi: 10.1038/s41467-023-41978-0
165. J. Fu et al., '19.31% binary organic solar cell and low non-radiative recombination enabled by non-monotonic intermediate state transition', *Nature Communications*, vol. 14, no. 1, p. 1760, Mar. 2023, doi: 10.1038/s41467-023-37526-5
166. N. Y. Doumon, L. Yang, and F. Rosei, 'Ternary organic solar cells: A review of the role of the third element', *Nano Energy*, vol. 94, p. 106915, Apr. 2022, doi: 10.1016/j.nanoen.2021.106915
167. Q. An, F. Zhang, J. Zhang, W. Tang, Z. Deng, and B. Hu, 'Versatile ternary organic solar cells: A critical review', *Energy & Environmental Science*, vol. 9, no. 2, pp. 281–322, 2016, doi: 10.1039/C5EE02641E
168. G. Yu, J. Gao, J. C. Hummelen, F. Wudl, and A. J. Heeger, 'Polymer photovoltaic cells: Enhanced efficiencies via a network of internal donor-acceptor heterojunctions', *Science (1979)*, vol. 270, no. 5243, pp. 1789–1791, Dec. 1995, doi: 10.1126/science.270.5243.1789
169. J. J. M. Halls et al., 'Efficient photodiodes from interpenetrating polymer networks', *Nature*, vol. 376, no. 6540, pp. 498–500, Aug. 1995, doi: 10.1038/376498a0
170. G. Li et al., 'High-efficiency solution processable polymer photovoltaic cells by self-organization of polymer blends', *Nature Materials*, vol. 4, no. 11, pp. 864–868, Nov. 2005, doi: 10.1038/nmat1500
171. J. You et al., 'A polymer tandem solar cell with 10.6% power conversion efficiency', *Nature Communications*, vol. 4, no. 1, p. 1446, Feb. 2013, doi: 10.1038/ncomms2411
172. Y. Cui et al., 'Single-junction organic photovoltaic cell with 19% efficiency', *Advanced Materials*, vol. 33, no. 41, Oct. 2021, doi: 10.1002/adma.202102420
173. J. Wang et al., 'Tandem organic solar cells with 20.6% efficiency enabled by reduced voltage losses', *National Science Review*, vol. 10, no. 6, May 2023, doi: 10.1093/nsr/nwad085
174. K. Liu, Y. Jiang, G. Ran, F. Liu, W. Zhang, and X. Zhu, '19.7% efficiency binary organic solar cells achieved by selective core fluorination of nonfullerene electron acceptors', *Joule*, vol. 8, no. 3, pp. 835–851, Mar. 2024, doi: 10.1016/j.joule.2024.01.005
175. Y. Lin et al., 'An electron acceptor challenging fullerenes for efficient polymer solar cells', *Advanced Materials*, vol. 27, no. 7, pp. 1170–1174, Feb. 2015, doi: 10.1002/adma.201404317
176. W. Ni, X. Wan, M. Li, Y. Wang, and Y. Chen, 'A–D–A small molecules for solution-processed organic photovoltaic cells', *Chemical Communications*, vol. 51, no. 24, pp. 4936–4950, 2015, doi: 10.1039/C4CC09758K
177. L. Xiao et al., 'A-D-A type small molecules based on boron dipyrromethene for solution-processed organic solar cells', *Chemistry: An Asian Journal*, vol. 10, no. 7, pp. 1513–1518, Jul. 2015, doi: 10.1002/asia.201500382

178. Z. Xiao et al., '26 mA cm^{-2} J_{sc} from organic solar cells with a low-bandgap nonfullerene acceptor', *Science Bulletin (Beijing)*, vol. 62, no. 22, pp. 1494–1496, Nov. 2017, doi: 10.1016/j.scib.2017.10.017
179. Y. Zhang, Y. Lang, and G. Li, 'Recent advances of non-fullerene organic solar cells: From materials and morphology to devices and applications', *EcoMat*, vol. 5, no. 1, Jan. 2023, doi: 10.1002/eom2.12281
180. J. Lv et al., 'Additive-induced miscibility regulation and hierarchical morphology enable 17.5% binary organic solar cells', *Energy & Environmental Science*, vol. 14, no. 5, pp. 3044–3052, 2021, doi: 10.1039/D0EE04012F
181. Y. Zhang et al., 'Graded bulk-heterojunction enables 17% binary organic solar cells via nonhalogenated open air coating', *Nature Communications*, vol. 12, no. 1, p. 4815, Aug. 2021, doi: 10.1038/s41467-021-25148-8
182. A. Shang et al., 'Over 18% binary organic solar cells enabled by isomerization of non-fullerene acceptors with alkylthiophene side chains', *Science China Chemistry*, vol. 65, no. 9, pp. 1758–1766, Sep. 2022, doi: 10.1007/s11426-022-1290-y
183. Y. Wei et al., 'Binary organic solar cells breaking 19% via manipulating the vertical component distribution', *Advanced Materials*, vol. 34, no. 33, Aug. 2022, doi: 10.1002/adma.202204718
184. X. Xu, K. Feng, Y. W. Lee, H. Y. Woo, G. Zhang, and Q. Peng, 'Subtle polymer donor and molecular acceptor design enable efficient polymer solar cells with a very small energy loss', *Advanced Functional Materials*, vol. 30, no. 9, Feb. 2020, doi: 10.1002/adfm.201907570
185. G. Xie, Z. Zhang, Z. Su, X. Zhang, and J. Zhang, '16.5% efficiency ternary organic photovoltaics with two polymer donors by optimizing molecular arrangement and phase separation', *Nano Energy*, vol. 69, p. 104447, Mar. 2020, doi: 10.1016/j.nanoen.2020.104447
186. J. Han, X. Wang, D. Huang, C. Yang, R. Yang, and X. Bao, 'Employing asymmetrical thieno[3,4-*d*]pyridazin-1(2 *H*)-one block enables efficient ternary polymer solar cells with improved light-harvesting and morphological properties', *Macromolecules*, vol. 53, no. 15, pp. 6619–6629, Aug. 2020, doi: 10.1021/acs.macromol.0c00459
187. Q. An et al., 'Two compatible polymer donors contribute synergistically for ternary organic solar cells with 17.53% efficiency', *Energy & Environmental Science*, vol. 13, no. 12, pp. 5039–5047, 2020, doi: 10.1039/D0EE02516J
188. K. Li et al., 'Ternary blended fullerene-free polymer solar cells with 16.5% efficiency enabled with a higher-LUMO-level acceptor to improve film morphology', *Advanced Energy Materials*, vol. 9, no. 33, Sep. 2019, doi: 10.1002/aenm.201901728
189. L. Zhan et al., 'Over 17% efficiency ternary organic solar cells enabled by two non-fullerene acceptors working in an alloy-like model', *Energy & Environmental Science*, vol. 13, no. 2, pp. 635–645, 2020, doi: 10.1039/C9EE03710A
190. C. Zhu et al., 'Tuning the electron-deficient core of a non-fullerene acceptor to achieve over 17% efficiency in a single-junction organic solar cell', *Energy & Environmental Science*, vol. 13, no. 8, pp. 2459–2466, 2020, doi: 10.1039/D0EE00862A
191. Y. Li et al., 'A facile strategy for third-component selection in non-fullerene acceptor-based ternary organic solar cells', *Energy & Environmental Science*, vol. 14, no. 9, pp. 5009–5016, 2021, doi: 10.1039/D1EE01864G
192. Y. Cai et al., 'A well-mixed phase formed by two compatible non-fullerene acceptors enables ternary organic solar cells with efficiency over 18.6%', *Advanced Materials*, vol. 33, no. 33, Aug. 2021, doi: 10.1002/adma.202101733
193. Z. Chen et al., 'Small-molecular donor guest achieves rigid 18.5% and flexible 15.9% efficiency organic photovoltaic via fine-tuning microstructure morphology', *Joule*, vol. 5, no. 9, pp. 2395–2407, Sep. 2021, doi: 10.1016/j.joule.2021.06.017

5 Synthesis and Fabrication of Multilayer Graded Bandgap Solar Cells

S.K. Tripathi, Mamta Goel, and Krupali Shetty*

5.1 INTRODUCTION

Shockley-Queisser limit, at 33.7%, indicates the maximum theoretical efficiency of a single junction solar cell operating at one sun [1]. The main reasons this limit is low are transmission and thermalization losses, which together make up around 56% of the total loss. These losses result from the use of a single energy band gap in order to convert a wide range of photons that come from solar energy. Thermalization losses are the result of all other high-energy photons being absorbed and transformed into heat above the band gap. Transmission loss results from red photons with energy below the band gap not being absorbed. Shockley-Queisser limit is not the ultimate limit one may attain with photovoltaic Thermodynamic concepts may be used to calculate the ultimate limit of solar energy, demonstrating that up to 86% of solar energy can be converted into electrical power [2]. Since a single material has just two energy levels, the Shockley-Queisser limit and the thermodynamic limit differ greatly, despite the fact that photons in the solar spectrum have a broad range of energies. It is inherently inappropriate for devices that rely on a single energy level transition to operate in the wide spectrum of sunlight. It is essential to make the most use of every photon while using the sun's spectrum. Figure 5.1a schematically depicts the photon efficiency spectrum of an optimized single junction (single band gap) solar cell. The colored area represents the photon efficiency spectrum, while the light grey area displays the black body spectrum of a body at 5760 K, representing the spectrum of the sun. Since the spectrum of the sun contains an infinite number of photons, so to attain overall high efficiency, there should be an infinite number of band gaps in order to convert each photon to almost perfect efficiency. A finite number of band gaps (2, 3, or more) can be used if there aren't any infinite band gaps. The tandem solar cell concept is based on the use of several band gaps to collect a larger range of the sun spectrum and boost overall efficiency. The main objective of the tandem solar cell is to take a considerable quantity of the sunlight spectrum in order to enhance the effectiveness of the gadget by raising the electric current. The method involves creating p–n, p–i–n, or other diode structures using semiconductors with varying bandgaps and linking them together to improve photovoltaic conversion. Figure 5.1b shows the arrangement of three separate p–i–n diodes to use photons in three areas.

 DOI: 10.1201/9781003499701-5

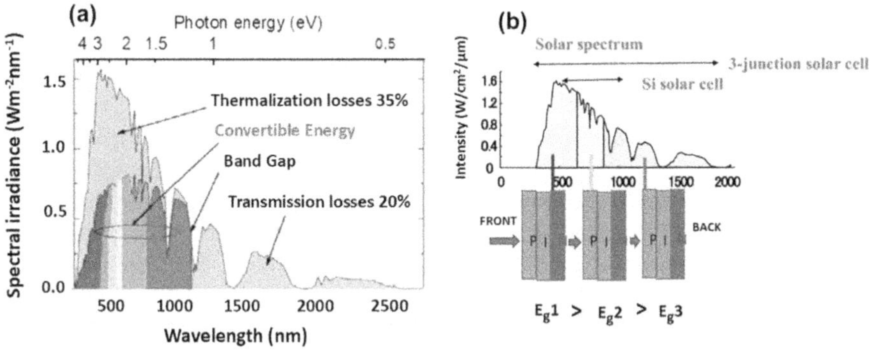

FIGURE 5.1 (a) Photon efficiency as a function of photon wavelength (b) arrangement of three separate p–i–n diodes to use photons in three areas. Typically, solar cells are stacked in descending order of band gap. (Adapted with permission from Elsevier through exclusive licensee [Elsevier and Springer].)

Typically, photovoltaic cells are stacked in descending order of band gap. To be able to convert high-energy photons at the blue end of the spectrum, the first cell at the front is constructed using a broad bandgap material. Conversely, the third cell at the rear is made using a tight bandgap material to convert low-energy photons at the infrared end of the spectrum. Following this methodology, the subsequent action involves interconnecting the individual cells. In 2024, classic crystalline silicon (c-Si) solar cells achieved an efficiency of up to 27.1% [3]. Similarly, laboratory samples of multi-junction cells or so-called Tandem solar cells have shown performance exceeding 46% when exposed to focused sunshine [3]. Tandem cells have become commercially accessible with an efficiency of 30% under one-sun illumination and can reach an efficiency of over 40% under concentrated sunlight. Under concentration, GaAs multijunction solar cells and similar structures have been produced with above 40% efficiency [4]. However, this enhanced efficiency comes with the drawback of heightened complexity and higher production costs.

Important devices have been revisited to explore various connections for multilayer tandem solar cells.

1. Silicon is the most common material used in photovoltaic applications; however, its constant band gap (1.1 eV) is the primary issue limiting the effectiveness of silicon-based solar cells. It has been figured out that in cases when the size of a Si crystal is lowered below a certain level, the material begins to act differently. The optical and electrical characteristics of all semiconductor materials alter. When the crystal Si is not as large as the Bohr exciton radius. As the nanoparticles' size shrinks and approaches the Bohr exciton radius, the lowest accessible energy state in conduction and the maximum occupied energy state in the valance band divide and move apart, increasing the effective band gap. The following equation illustrates how the band gap alters as the size of the nanocrystal decreases:

$$E_g = E_g(\text{bulk}) + \frac{k}{d^2} \tag{5.1}$$

Where k is the quantum confinement parameter, d is nanoparticle diameter in nanometers, and E_g (bulk) is the bulk material's band gap. Tiny nanocrystals that resemble the Bohr radius are called Quantum dots (QD). By adjusting the size of Si quantum dots, a band gap as high as 2.7 eV may be produced [4,5]. To establish a multijunction solar cell, a wide band of silicon material might be coupled with bulk crystalline silicon. For instance, the bottom cell of a tandem solar cell arrangement might be a crystalline silicon wafer with a band gap of 1.12 eV, while an artificial material based on silicon with a large band gap can function as the top cell. There are chances for adjusting the band gap of this artificial material to obtain the highest possible effectiveness of the multijunction solar cell. In a similar manner, the bulk crystalline silicon may be coupled with two artificial wide band gap silicon-based materials that have distinct band gaps in order to produce all-silicon triple junction solar cells.

2. Another idea for employing sub-bandgap photons to boost solar cell efficiency is the impurity photovoltaic (IPV) effect. This effect also increases the effectiveness of solar cells in a similar manner. This is accomplished by utilizing two-step production via impurity states that are present inside the band gap. By employing the three transitions that are accessible inside the system, a more accurate matching of the sun's spectrum may be achieved. IPV photovoltaic cells are expected to achieve high efficiency, according to predictions. A substance with a wide band gap may be utilized in the IPV scenario. The problem here is to identify a host material that is acceptable and a material that is suitable to supply intermediate energy levels. These energy levels are useful just for the production of carriers, but they do not promote the recombination of carriers. Within the framework of this approach, it is possible to forecast efficiency of up to 63% [6].
3. The intermediate-band (IB) photovoltaic cell is intended to offer a large photogenerated current while retaining a high output voltage. To make this conceivable, these cells integrate an energy band that is partly filled with electrons inside the forbidden bandgap of a semiconductor. This intermediate band may be used as a stepping stone by photons lacking the energy to force electrons between the valence band and the conduction band, forming an electron–hole pair. Nanostructured materials and certain alloys have been exploited in the practical implementation of intermediate-band solar cells. However, hurdles still exist in producing practical systems. The limiting effectiveness of stacked solar cells (three) and the limiting effectiveness of single impurity level and single intermediate band gap are both above 63% [7,8]. The benefit of IPV and IB concepts is that the same material is used throughout, and interconnection between cells occurs automatically.
4. There are two possible approaches for creating a tandem solar cell: joining them in series or in parallel arrangements. As a practice, tandem solar cell uses series connections, where the tunnel junction connects adjacent

devices. The design shows a schematic design with n-type material linked to neighboring p-type material (Figure 5.2a). The objective is to convert photons with diverse energy ranges in three devices and effectively collect charge carriers to create high power in the external circuit. One device's valence band and the next's conduction band are connected.

Multilayer graded bandgap solar cell is a device that combines multiple devices in parallel, changing their conduction type from n⁺ to p⁺ through a sequence of layers [9]. This process can be repeated for p-type materials and may be split into numerous separate solar cells. The conduction band of One device's conduction band is connected to the adjacent device's conduction band, resulting in parallel connections (Figure 5.2b). The bandgap of the solar cell can be gradually lowered from the front to the back in order to achieve full absorption of the solar spectrum. However, since the device's open-circuit voltage depends on the smallest bandgap material utilized in its construction, the lowest bandgap material is forfeited. It has been found that the current method of using a series connection or tunnel junction technique is adverse to the arrangement of charge providers in tandem cells. Through the basics of each tunnel junction into the structure of the photovoltaic cell, the charge carriers are slowed down, and they are given the opportunity to recombine in the area of the tunnel

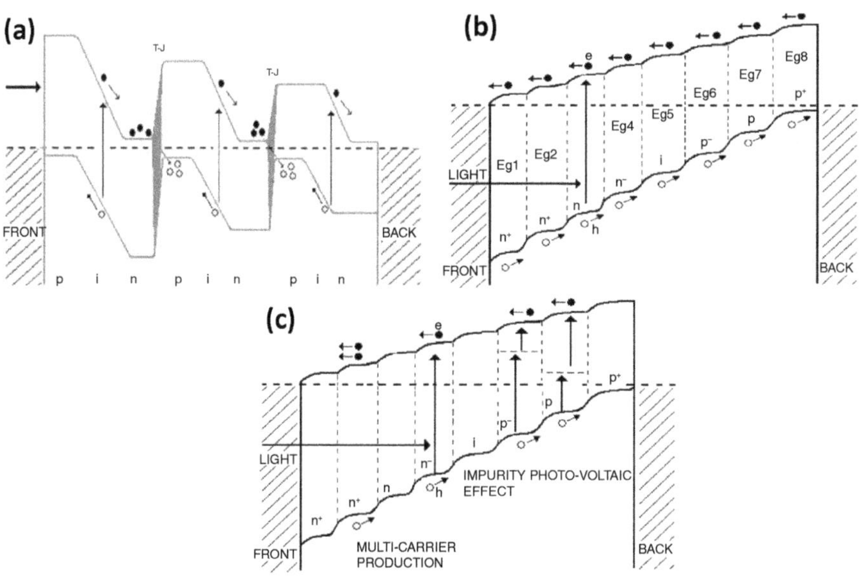

FIGURE 5.2 Schematic design (a) with n-type material linked to neighboring p-type material tandem solar cell fabricated by connecting three p–i–n diodes in series through two tunnel junctions (b) a multilayer graded band gap solar cell, formed by connecting many cells (eight cells here) in parallel. The material layers' band gap progressively decreases, and the cell's conduction type varies between n⁺ and p⁺ from front to rear. (c) The potential for multiple charge carrier generation via the avalanche effect and the multiple-step generation of charge carriers with IR photons via the IPV effect in tandem cells linked in parallel. (Adapted with permission from Elsevier through exclusive licensee [Elsevier].)

junction. Multi-layer graded bandgap solar cells are created by connecting multiple thin film solar cells in parallel form. The adoption of parallel connections offers various benefits over series connections, and there is huge potential for generating higher efficiencies in third-generation tandem photovoltaic cells. The carriers of charges in these devices are continually accelerated over the whole of the device and they're collected in the external circuit. Multi charge-carrier manufacturing and impurity photo-voltaic mechanisms are also included in this system to boost its performance by raising the short-circuit current density.

5.2 NEED AND IMPORTANCE OF MULTILAYER GRADED BANDGAP SOLAR CELLS

1. By utilizing multiple layers with varying bandgaps, these cells can capture a broader range of sunlight wavelengths.
2. Lower thermalization loss as a result of higher photon absorption in various cell regions.
3. The reduction of recombination and generation (R&G) by the enhancement of e-h pair collection caused by the existence of an electric field (or depletion width) that roughly extends the whole thickness of the photovoltaic cell [7].
4. IPV effect and impact ionization are included in the design of multi-layer graded band gap solar cells to lower recombination and boost photo-generated current density even further [9].

Tandem solar cells coupled with tunnel connections have some drawbacks. Due to an internal electric field operating in opposite directions, the photogenerated electrons of one cell are drawn closer to the photogenerated holes of the next cell, slowing the carriers and increasing the rate of carrier recombination. Actually, if all three solar cells absorb the same quantity of photons, only holes from the front cell and electrons from the rear cell will be experienced by the external circuit. In this kind of situation, having more cells is not particularly beneficial. As the number of cells increases even higher, costs rise with such a technique, while efficiency increases are negligible.

While in parallel connections as given in Figure 5.2c. The cell uses high-energy photons in its front and low-energy photons in its rear layers to produce electron-hole pairs, with electrons directed toward the front contact and holes toward the rear contact. The constant acceleration minimizes recombination, allowing charge carriers to reach high kinetic energies, permitting each charge carrier created by light absorption to go to the external circuit load. As the charge carriers experience continual acceleration, they may potentially form numerous charge carriers owing to an avalanche effect boosting the electric current in the device.

Furthermore, devices coupled in parallel also include the intermediate band photovoltaic effect or IPV effect. Numerous e-h pairs might be produced by various combinations of IR photons employing varying degrees of impurity present in the semiconductor layers, ranging from front to rear. As given in Figure 5.2c, this impact may also enhance the PV activity and current collection from this kind of device construction.

5.3 FABRICATION AND SYNTHESIS OF MULTILAYER GRADED BANDGAP SOLAR CELLS MATERIALS

Semiconductor materials from the II–VI group for photovoltaic cells have shown provided great promise in recent years. These materials, like cadmium telluride (CdTe), Zinc Sulphide (ZnS), Cadmium Sulphide (CdS), have strong light-absorption capabilities due to direct band gap nature and can be easily mass-produced at a low cost. This makes them a desirable choice for large-scale generation of renewable energy. Additionally, continuous investigation and advancement in this area continue to improve the effectiveness and reliability of II-VI semiconductors for use in photovoltaic technology.

Efficient CdTe-based solar cells have seen slow progress in the last 20 years, with a record efficiency of 22% in 2018 [8]. To improve efficiency, suggestions include reducing CdS window layer thickness, using new window materials, buffer layers, back contact materials, and emerging design structures like multi-layer graded-bandgap and Schottky contact. The thickness of the CdS layer has been suggested as a way to increase the efficiency of CdS/CdTe solar cells due to its high absorption coefficient. The thick CdS layers absorb a significant quantity of light in the visible region of the sun's spectrum, preventing charge carriers from obtaining the CdTe absorber material. This results in a decrease in short-circuit current density that allows more photons to reach the absorber material to charge carrier generation and collection. However, this idea faces challenges, resulting in pinholes and reduction of fill factors and open circuit voltage. To overcome these issues, buffer layers such as ZnO, Al-ZnO, SnO_2, or ZTO can be brought into use [10,11]. The use of a buffer layer reduces short circuits but may limit short-circuit current density due to additional series resistance. A rough surface of TCO contact can cause shunting problems on decreasing the CdS layer when CdS. A multi-layer graded-bandgap solar cell approach is used for producing a CdTe solar cell, improving the efficiency of the device using electrodeposited n-CdS, N-CdTe, and ZnS as discussed in the earlier section. The resulting interfaces support a strong internal electric field. One of the reports used a two-electrode method to grow three semiconductors in spite of traditional three electrodes due to the potential for ions like Ag^+ and Na^+ to reduce conversion efficiency in CdS/CdTe solar cells. The existence of ions such as Ag and K in the CdTe solution may lead to cell deterioration.

5.3.1 Importance and Synthesis of ZnS for Solar Cell

Due to its superior electrical and optical characteristics, ZnS from II–VI compound semiconductors is an essential material utilized in the manufacturing of solar cells. Furthermore, ZnS can also be used as a buffer layer to clear the interface between different layers in the solar cell, reducing losses and increasing overall efficiency [12]. Its stability and compatibility with other materials make it a valuable component in the production of high-quality solar panels. In addition, ZnS has a long lifespan and is resistant to degradation from environmental factors, ensuring the longevity of solar panels. Its low cost and abundance also make it a cheap in price option for mass production of solar cells, helping to get green energy technology widely used. High photovoltaic conversion efficiency is possible using CdS as a window layer in CdTe-based photovoltaic cells, but its 2.42 eV bandgap restricts its application [13]. The wide bandgap (~3.70 eV) and excellent transparency of ZnS make it a promising alternative to CdS for improving solar panel efficiency and performance [14]. However, ZnS's high lattice mismatch, high conduction band edge, and high resistivity need CdS to be retained in the ZnS/CdS/CdTe solar cell arrangement. Significantly, ZnS offers excellent wetting and decreases CdS thickness without increasing shunting pathways [15] and parasitic absorption simplification.

Various methods, including spray pyrolysis [16], electrode position [17], and metal-organic chemical vapor deposition [18], have been documented for the growth of ZnS. Benefits from electrodeposition include cheap cost, ease of use, scalability, and high surface area manufacturing. It is accomplished by adding complex substances to the electrolytic bath, such as glycerol, hydrogen, ammonia, and trisodium citrate.

Synthesis of ZnS without using the complexing agents for multi-layer graded band gap photovoltaic Cells is reported by Ojo et al. [19]. ZnS fine films are electrodeposited on glass/FTO materials using a potentiostatic technique with a high-purity graphite rod serving as the counter electrode. Precursors included ammonium thiosulphate ($(NH_4)_2S_2O_3$) and zinc sulfate monohydrate ($ZnSO_4H_2O$). In a 500 mL polypropylene beaker, 400 mL of deionized water was used to dissolve 0.2 M $ZnSO_4H_2O$ and 0.2 M $(NH_4)_2S_2O_3$ to create the electrolytic bath. To accomplish total dissolution and decrease impurity, the solution was constantly agitated and electro-purified at 1000 mV for about 100 hours. ZnS layers with a pH of between 3.00 and 4.50 were characterized at various cathodic voltages. The working and counter electrodes in the two-electrode electrodeposition setup were a high-purity carbon electrode and a glass/FTO substrate. The obtained electrolyte(s)'s cyclic voltammograms were employed to measure the potential ZnS deposition voltage range. The n-ZnS was air-cooled, heat-treated at 300°C, and then washed before the CdS was deposited. In the next session, the process for depositing the n-CdS/n-CdTe layers for device designs is covered.

5.3.2 Role of CdS Material and Its Synthesis

Because of its adjustable bandgaps and high absorption coefficients, CdS has attracted a lot of interest for use in photovoltaic applications. These substances show promise as low-cost and effective energy conversion materials for thin-film solar cells. To improve conversion efficiency, polycrystalline CdS fine films are frequently used as a layer of the window in CdS/CdTe photovoltaic cell combinations [20]. Furthermore, CdS and CdSe are perfect for long-term usage in solar panels because of their exceptional stability and longevity under challenging climatic circumstances. Additionally, because they operate on flexible substrates, this opens up possibilities for solar energy systems that are lightweight and flexible. Overall, the combination of high efficiency, durability, and flexibility makes CdS the ideal material for the progress of next-generation solar technologies. CdS fine films have been produced by a variety of techniques, including chemical vapor deposition, thermal evaporation, electrodeposition, and spray pyrolysis. In terms of film quality, cost-effectiveness, and scalability, each technique has certain benefits. The particular requirements of the solar cell design and the intended performance parameters determine which deposition process is used. CdS films are electrodeposited by using sodium thiosulfate ($Na_2S_2O_3$) as the sulfur precursor [15]. The further development of CdS from sulfur precursors like thiourea (NH_2CSNH_2) is an encouraging method because of its capacity to deposit CdS without the drawbacks of p-type dopants like Na. However, the electrical conductance of the grown layer is reduced by these dopants. The electrodeposition technique is advantageous for deposition process continuity and Cd-containing waste reduction. CdS thin films can be electrodeposited using an electrolytic bath, containing 0.12 M $CdCl_2{\cdot}xH_2O$ (Cadmium chloride hydrate) and 0.18 M NH_2CSNH_2 as cadmium and sulfur precursor in 800 mL DI water in a 1000 mL polypropylene beaker, which was placed in an 1800 mL glass beaker containing DI water. The solution was stirred and electro-purified for approximately 50 hours to reduce impurity levels and achieve homogeneity.

5.3.3 Synthesis of CdTe Material for Multi-layer Graded Bandgap Solar Cells

CdTe is among the most prominent of the group, and its characteristics have been briefly explored [21,22]. As demonstrated in the literature, CdTe has a broad variety of uses for optoelectronic devices owing to its near direct bandgap of 1.45 eV of the high absorption coefficient of roughly $10^4 cm^{-1}$ within the solar spectrum's visible and infrared regions of the sun's spectrum at 2 μm thickness. The hunt for low-cost methods for mass manufacturing and high optoelectronic conversion effectiveness of photovoltaic solar cells has spurred the employment of diverse thin-film growth processes in the deposition of CdTe. With over 14 distinct growth strategies published in the work of literature, economical, simplicity, scalability, and fabrication have been a few features of the electrodeposition approach beyond others [23]. Some common precursors for CdTe are $Cd(NO_3)_2$ and $CdCl_2$, $CdSO_4$.

CdTe fine films can be catholically electrodeposited on FTO or glass materials with a potentiostatic method [24]. Cadmium nitrate tetrahydrate ($Cd(NO_3)_2$-$4H_2O$) was employed as the cadmium precursor, while Tellurium Oxide (TeO_2) served as the tellurium source. The aqueous electrolyte was made by dissolving 1.5 M $Cd(NO_3)_2$-$4H_2O$ in 800 mL of deionized water and placing it in an 1800 mL glass beaker. The $Cd(NO_3)$-$4H_2O$ solution was electropurified for 50 hours before adding 0.0002 M TeO_2 and stirring for 5 hours to achieve homogeneity. To solve solubility difficulties, a 0.03 M TeO_2 solution was created by dissolving it in strong nitric acid (HNO_3) and stirring for 1 hour.

5.4 MULTI-LAYER GRADED BANDGAP SOLAR CELLS FABRICATION AND CHARACTERIZATION AT THE LAB LEVEL

To create ZnS, CdS, and CdTe layers in multilayer graded band gap solar cell configuration on FTO substrate, two-electrode method can be used as discussed in Sections 5.3.1–5.3.3. The substrate is cleaned by using soap, acetone, and methanol, and de-ionized water rinsed between chemicals. High-purity graphite rod was used as an anode. The cathodic potential used to deposit ZnS, CdS, CdTe layers are

1550, 1450, and 1840 mV. The double-layered glass/FTO/CdS/CdTe configuration and triple-layered glass/FTO/n-ZnS/n-CdS/n-CdTe configuration (Figure 5.3a) of multi-layer graded band gap solar cell are etched in aqueous solutions, rinsed, and dried throughout the construction process. Vacuum-evaporated Gold (Au) metal contacts of 2 mm diameter and 100 nm thickness are used to form Schottky barriers on etched CdTe surfaces. The thickness of the ZnS layer and CdTe layer is 290 nm and 1.70 μm respectively. The thickness of the ZnS/CdS bi-layer is approximately 590 nm. According to reports, the cathodic potential to deposit the ZnS layer should be between 1350 and 1450 mV. Crystalline ZnS can be deposited only between the limited voltage range of 1420–1430 mV. As the voltage goes above 1425 mV, P-type ZnS changes into N-type ZnS.

As confirmed by the cyclic voltammogram the Te is deposited at a cathodic voltage of around 220 mV. Figure 5.3b shows that the deposition of Cd starts at point A while the CdTe starts at point B.

Past this stage, water electrolysis starts and Cd-rich CdTe develops. Reverse swipe shows that Te is stripped from the substrate more quickly than Cd is from 670 mV and below. This yields a CdTe layer that is very Te-rich and is precisely electrodeposited to promote n-type conductivity and enhance solar cell efficiency.

Conduction type (n or p) of various semiconductor layers of the cell can be confirmed by using Photo electrochemical cell measurements. Band gap information of various layers can be obtained from UV-Vis spectrophotometers. The orientation of

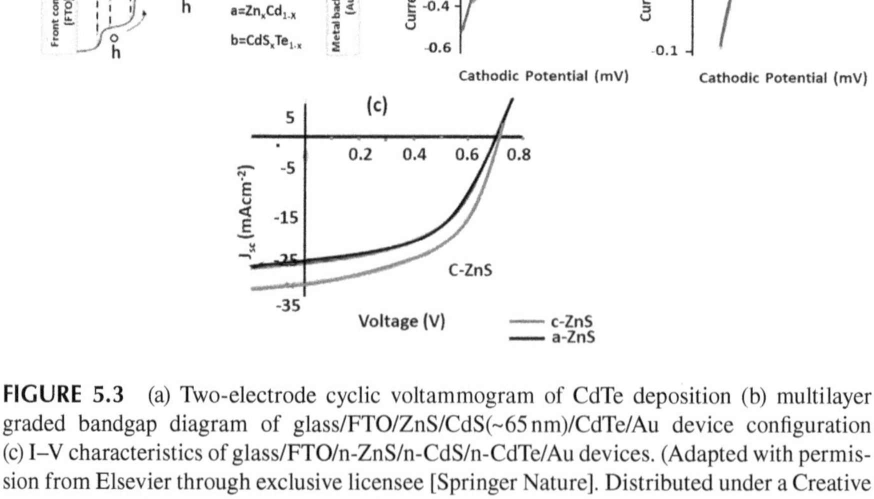

FIGURE 5.3 (a) Two-electrode cyclic voltammogram of CdTe deposition (b) multilayer graded bandgap diagram of glass/FTO/ZnS/CdS(~65 nm)/CdTe/Au device configuration (c) I–V characteristics of glass/FTO/n-ZnS/n-CdS/n-CdTe/Au devices. (Adapted with permission from Elsevier through exclusive licensee [Springer Nature]. Distributed under a Creative Commons Attribution License 4.0 (CC BY) https://creativecommons.org/licenses/by/4.0/)

crystallite can be found by using X-Ray Diffractograms. ZnS films that have been deposited have tiny grains with band gaps between 3.70 and 3.90 eV. Various parameters of multilayer graded band gap solar cells for four different configurations have been summarized in Table 5.1. Compared to devices made with an amorphous ZnS buffer layer, solar cells produced with crystalline ZnS have better solar cell properties. Also, triple-layer multi-layer graded solar cells using the ZnS layer have shown large efficiency and high charge carrier concentration as compared to double-layer solar cells without ZnS, which clearly indicates the advantages of using multilayer graded band gap configuration.

TABLE 5.1
Results of Various Characterizations for Few Multilayer Graded Band Gaps Solar Cells for Few Different Structures [19,25]

Characterization	Device Parameters	Crystalline ZnS Based Glass/FTO/ZnS/CdS/CdTe/Au Multilayer Graded Band Gap Solar Cell	AmorphousZnS Based Glass/FTO/ZnS/CdS/CdTe/Au Multilayer Graded Band Gap Solar Cell	Triple Layer Glass/FTO/ZnS/CdS/CdTe/Au 3-Layer Multilayer Graded Band Gap Solar Cell	Double Layered Glass/FTO/CdS/CdTe solar Cell
I–V measurement under dark condition	Shunt resistance, R_{sh} (Ω)	$>10^6$	$>10^6$	3825	4574
	Series resistance, R_s (Ω)	470	637	175	316
	log (R.F.)	4.8	4.2	4.4	4.1
	I_o (A)	1.0×10^{-9}	1.2×10^{-9}	2.5×10^{-9}	3.1×10^{-9}
	Ideality factor n	1.60	1.85	2.37	2.54
	Barrier potential, Φ_b (eV)	>0.82	>0.79	>1.13	>1.10
I–V measurement under AM1.5 illuminated condition	Short circuit current I_{SC} (mA)	1.07	0.94	1.28	1.21
	Short circuit current density, J_{SC} (mA/cm^2)	34.08	29.94	40.8	38.5
	Open circuit voltage, V_{OC} (mV)	730	720	640	630
	Fill-factor	0.57	0.52	0.40	0.33
	Efficiency (%)	14.18	11.21	10.4	8.0

(Continued)

TABLE 5.1 (*Continued*)
Results of Various Characterizations for Few Multilayer Graded Band Gaps Solar Cells for Few Different Structures [19,25]

Characterization	Device Parameters	Crystalline ZnS Based Glass/FTO/ZnS/CdS/CdTe/Au Multilayer Graded Band Gap Solar Cell	AmorphousZnS Based Glass/FTO/ZnS/CdS/CdTe/Au Multilayer Graded Band Gap Solar Cell	Triple Layer Glass/FTO/ZnS/CdS/CdTe/Au 3-Layer Multilayer Graded Band Gap Solar Cell	Double Layered Glass/FTO/CdS/CdTe solar Cell
DC and C–V measurements under dark condition	Resistivity, $\rho \times 10^3$ (Ωcm)	1.13	1.13		
	Capacitance, C_o (Pf)	280	340	151	179
	Barrier potential V_{bi} (eV)	1.10	1.07		
	Doping concentration $(N_D - N_A)$ (cm^{-3})	7.79×10^{14}	9.52×10^{14}	8.1×10^{14}	2.5×10^{15}
	$(E_C - E_F)$ eV	~0.17	~0.21	—	—
	Depletion width W (nm)	1092.2	899.5	—	—
	Charge carrier mobility (cm^2/V s)	7.07	5.78	—	—

5.5 CHALLENGES IN THE FABRICATION OF MULTILAYER GRADED BANDGAP SOLAR CELL

The primary obstacles encountered in the production of multilayer graded band gap solar cells involve managing the electrodeposition process to account for thicker deposition layers, controlling the ionic concentration without the ability to measure it accurately, and addressing the non-uniformity of electroplated semiconductor layers caused by the nucleation process. Despite the high short circuit current obtained using multilayer configuration, the biggest challenge is the stability and reproducibility of these devices because of the complexity involved in fabrication.

5.6 CONCLUSIONS

This chapter describes the need, importance, and fabrication of multilayer graded band gap solar cells. The graded bandgap enhances the potential to capture a greater number of photons from both the blue and red regions of the solar spectrum. Tandem solar cells employing a series arrangement via tunnel junctions are inefficient for charge carrier collection since photo-generated electrons and holes decelerate and recombine. Multi-layer tandem solar cells with a parallel structure provide benefits, such as altering conduction type and bandgap, efficiently segregating, accelerating, and collecting photo-generated charge carriers. The key semiconductor materials studied and described in this chapter are ZnS, CdS, and CdTe. ZnS is the main buffer layer material. CdSe acts as the main window layer, and the function of the CdTe layer is to act like the main absorber layer. The chapter also describes the synthesis of these materials by using low-cost, scalable, and manufacturable electrodeposition techniques. The fabrication of double-layer glass/FTO/CdS/CdTe/Au and triple-layer glass/FTO/ZnS/CdS/CdTe multilayer graded band gap solar cell using amorphous and crystalline ZnS has been demonstrated. Even though solar cell parameters like high Jsc devices are frequently observed to be improving, these devices' poor stability and reproducibility remain the main problems.

NOTE

* Corresponding author: S.K. Tripathi, Center of Advanced Study in Physics, Department of Physics, Panjab University, Chandigarh, India.

REFERENCES

1. Shockley W, Queisser HJ. Detailed balance limit of efficiency of p-n junction solar cells. *Journal of Applied Physics*. 1961 Mar 1;32:510–9.
2. Nelson J. *The Physics of Solar Cells* [Internet]. 2003rd ed. Vol. 2003. UK: Imperial College Press; Available from: doi.org/10.1142/p276
3. Dimroth F, Tibbits TND, Niemeyer M, Predan F, Beutel P, Karcher C, et al. Four-junction wafer-bonded concentrator solar cells. *IEEE Journal of Photovoltaics*. 2016 Jan;6(1):343–9.
4. Ramadan R, Manso-Silván M, Martín-Palma RJ. Hybrid porous silicon/silver nanostructures for the development of enhanced photovoltaic devices. *Journal of Materials Science*. 2020 May 1;55(13):5458–70.
5. Selopal GS, Zhao H, Wang ZM, Rosei F. Core/shell quantum dots solar cells. *Advanced Functional Materials*. 2020;30(13):1908762.
6. Khan Md SP, Farzana E. Impurity photovoltaic effect in multijunction solar cells. *Procedia Technology*. 2013 Jan 1;7:166–72.
7. Dharmadasa IM, Elsherif O, Tolan GJ. Solar cells active in complete darkness. *Journal of Physics: Conference Series*. 2011 Mar;286(1):012041.
8. ScienceDirect Topics. Cadmium Telluride Solar Cell - An Overview. [Internet]. [cited 2024 Apr 21]. Available from: https://www.sciencedirect.com/topics/engineering/cadmium-telluride-solar-cell
9. Dharmadasa IM, Ojo AA, Salim HI, Dharmadasa R. Next generation solar cells based on graded bandgap device structures utilising rod-type nano-materials. *Energies*. 2015 Jun;8(6):5440–58.
10. Rani M, Tripathi SK. Electron transfer properties of organic dye sensitized ZnO and ZnO/TiO_2 photoanode for dye sensitized solar cells. *Renewable and Sustainable Energy Reviews*. 2016 Aug 1;61:97–107.
11. Rani M, Tripathi SK. Effect of Eosin Y Dye on electrical properties of ZnO film synthesized by sol–gel technique. *Journal of Electronic Materials*. 2014 Feb;43(2):426–34.
12. Nakada T, Mizutani M. 18% Efficiency Cd-free Cu(In, Ga)Se_2 thin-film solar cells fabricated using chemical bath deposition (CBD)-ZnS buffer layers. *Japanese Journal of Applied Physics*. 2002 Feb 1;41(2B):L165.

13. Britt J, Ferekides C. Thin-film CdS/CdTe solar cell with 15.8% efficiency. *Applied Physics Letters.* 1993 May 31;62(22):2851–2.
14. Ennaoui A, Eisele W, Lux-Steiner M, Niesen TP, Karg F. Highly efficient Cu(Ga,In)$(S,Se)_2$ thin film solar cells with zinc-compound buffer layers. *Thin Solid Films.* 2003 May 1;431–432:335–9.
15. Ojo AA, Salim HI, Olusola OI, Madugu ML, Dharmadasa IM. Effect of thickness: A case study of electrodeposited CdS in CdS/CdTe based photovoltaic devices. *Journal of Materials Science: Materials in Electronics.* 2017 Feb 1;28(4):3254–63.
16. Al-Diabat AM, Ahmed NM, Hashim MR, Almessiere MA. Growth of ZnS thin films using chemical spray pyrolysis technique. *Materials Today: Proceedings.* 2019 Jan 1;17:912–20.
17. Boosagulla D, Mandati S, Allikayala R, Sarada BV. Growth mechanism of pulse electrodeposited cadmium sulfide and zinc sulfide thin films with tartaric acid and glycerol as additives. *Thin Solid Films.* 2022 Jan 1;741:139011.
18. Palve AM. Deposition of zinc sulfide thin films from zinc(II) thiosemicarbazones as single molecular precursors using aerosol assisted chemical vapor deposition technique. *Frontiers in Materials* [Internet]. 2019 Mar 21 [cited 2024 Apr 21];6:46. Available from: https://www.frontiersin.org/articles/10.3389/fmats.2019.00046
19. Ojo AA, Salim HI, Dharmadasa I. The influence of ZnS crystallinity on all-electroplated ZnS/CdS/CdTe graded bandgap device properties. *Journal of Materials Science: Materials in Electronics* [Internet]. [cited 2024 Apr 21]. Available from: https://link.springer.com/article/10.1007/s10854-018-9491-4
20. Kapadnis RS, Bansode SB, Supekar AT, Bhujbal PK, Kale SS, Jadkar SR, et al. Cadmium telluride/cadmium sulfide thin films solar cells: A review. *ES Energy & Environment.* 2020;10:3–12. [cited 2024 Apr 21]. Available from: https://www.espublisher.com/journals/articlehtml/es-energy-environment/10.30919-esee8c706
21. Scarpulla MA, McCandless B, Phillips AB, Yan Y, Heben MJ, Wolden C, et al. CdTe-based thin film photovoltaics: Recent advances, current challenges and future prospects. *Solar Energy Materials and Solar Cells.* 2023 Jun 15;255:112289.
22. Photovoltaic Research. NREL. Cadmium Telluride Solar Cells. [Internet]. [cited 2024 Apr 21]. Available from: https://www.nrel.gov/pv/cadmium-telluride-solar-cells.html
23. Ling J, Zhang X, Mao T, Li L, Wang S, Cao M, et al. Electrodeposition of CdTe thin films for solar energy water splitting. *Materials (Basel).* 2020 Mar 27;13(7):1536.
24. Ojo AA, Dharmadasa IM. Analysis of the electronic properties of all-electroplated ZnS, CdS and CdTe graded bandgap photovoltaic device configuration. *Solar Energy.* 2017 Dec 1;158:721–7.
25. Echendu OK, Dharmadasa IM. Graded-bandgap solar cells using all-electrodeposited ZnS, CdS and CdTe thin-films. *Energies.* 2015 May;8(5):4416–35.

6 Solar Cell Market Growth and Their Applications

*Abhinav Nag, Anil Astakala, Shilpa Setia, Naveen Kumar, Shivani Dhall, and Summit Jalota**

6.1 INTRODUCTION

In the last two decades, photovoltaic (PV) installations have undergone remarkable growth, driven by technological advancements, cost reductions, and increased awareness of renewable energy benefits [1,2]. The surge in PV adoption, predominantly led by silicon-based solar cells, underscores the importance of understanding the silicon solar cell manufacturing process, its challenges, and recent advancements [3]. The effectiveness to combat the leading cause of climate change production and consumption, responsible for approximately 60% of greenhouse gas emissions rapid and efficient transition to renewable energy sources is imperative [4,5]. The escalation of global energy demands, driven by economic growth and rising incomes, underscores the pressing need for the swift expansion of renewable energy production. Geothermal, hydro, wind, and solar energy emerge as pivotal players in this transition, offering clean, sustainable, and abundant alternatives devoid of harmful emissions [6]. The substantial benefits of this shift towards solar energy include not only improved public health and heightened energy security but also a substantial reduction in greenhouse gas emissions [7]. Governments, businesses, and individuals must collaborate in creating an enabling environment that fosters the flourishing of renewable energy technologies. Furthermore, the sustained emphasis on developing markets for renewable energy sources is vital for nurturing innovation and competencies in the sustainable energy sector. First-generation solar cells primarily utilize monocrystalline or polycrystalline silicon, offering efficiencies of up to 24% and 13%–20%, respectively [7,8]. Monocrystalline silicon cells boast high efficiency and longevity but come at a higher cost, while polycrystalline silicon cells offer lower costs but slightly lower efficiency [9]. Second-generation options, including amorphous silicon, cadmium telluride, and copper indium gallium diselenide, offer lower costs and flexibility but generally have shorter lifetimes and lower efficiencies compared to first-generation cells [10]. Third-generation technologies, such as organic PV and concentrated PV, aim to overcome some of the limitations of earlier generations. Organic PV is lightweight and eco-friendly but has lower efficiency and shorter lifetimes [11]. Concentrated PV achieves very high efficiencies of up to 40% but comes with very high costs and requires integration with solar tracking systems and cooling devices to maintain peak performance [12]. Each generation of solar cells presents a trade-off between efficiency, cost, flexibility, and environmental impact, catering to different applications and priorities in the solar energy industry [13,14].

DOI: 10.1201/9781003499701-6

6.2 CURRENT CHALLENGES AT THE NATIONAL/INTERNATIONAL LEVEL

India faces critical challenges such as a poor supply chain of production and over-dependence on Chinese imports for solar cells [15]. India's abundant solar energy potential provides a clean and attainable replacement for conventional energy sources. However, the development of policies for viable utilization and technological advancement is crucial [16]. Academic institutions contribute significantly to solar cell research in India, with a focus on domestic and international collaborative research involving larger team sizes and materials other than silicon [17]. India aims to enhance solar manufacturing across the value chain and achieve self-reliance in solar cells and modules. However, concerns about economic feasibility and the impact on the grid and environment have been raised [18]. Jawaharlal Nehru National Solar Mission (JNNSM) aims to install a significant solar power capacity, but achieving this goal is challenging due to various barriers and challenges, including the need for a reliable direct normal irradiance database and indigenous manufacturing for concentrated solar power (CSP) plants [19].

In the world scenario, silicon-based solar cells, constituting over 90% of the commercial market, remain pivotal in the solar energy landscape [20]. With a focus on silicon solar cell manufacturing, this work aims to provide an overview of the value chain, incorporating cost and sustainability considerations [21]. From feedstock production to ingot processing to solar cell fabrication, each stage contributes to the efficiency and viability of solar energy production [22]. The journey begins with the raw materials required for silicon production, primarily quartz (SiO_2) and C-bearing compounds [2]. Metallurgical-grade silicon (MG-Si), with 98% purity, is initially obtained through processes like submerged arc furnaces or carbothermic reduction. Further refinement steps, such as slag- and vacuum refining, elevate the purity to solar-grade (SoG) silicon, exceeding 99.9999% [24]. Directional solidification (DS) plays a crucial role in this process by effectively rejecting metallic impurities, thereby enhancing silicon purity [25].

In silicon ingot production, monocrystalline ingots utilize the Czochralski (Cz) process, while directional solidification processes cater to multi-crystalline ingots [26]. The choice between these methods impacts the properties and performance of resulting solar cells. Key challenges in ingot production include optimizing yield, reducing costs, and enhancing efficiency to meet escalating demand [27].

Solar cell fabrication involves cutting wafers from silicon ingots and processing them into functional cells. This process includes saw damage removal, texturing, dopant diffusion to create a p-n junction, and the deposition of antireflection coating (ARC) and passivation layers [28]. Aluminum back surface field (Al-BSF) cells were prevalent until the emergence of the passivated emitter and rear cell (PERC) technology, which significantly enhances efficiency through additional passivation and reduced back surface field [29]. Variations like passivated emitter, rear locally diffused (PERL), and passivated emitter, rear diffused (PERT) further optimize performance. Beyond PERC, research has shifted towards tunnel oxide passivated contact (TOPCon) and silicon heterojunction (SHJ) cells, both featuring carrier-selective passivating contacts (CSPCs). These architectures address common challenges like Auger recombination and bandgap narrowing, achieving efficiencies upwards of 26.4% [30].

Recycling and sustainability aspects are also integral to the silicon solar cell value chain. Efforts to collect, disassemble, and process PV module waste aim to recover key materials like silicon, aluminum, glass, silver, and copper [31]. The solar cell industry faces multifaceted challenges at both national and international levels, shaping its trajectory and potential impact on global energy transitions. One of the foremost challenges lies in the realm of policy and regulation, where inconsistent or inadequate frameworks can hinder the growth of solar energy deployment [32]. Uncertainties surrounding government incentives, subsidies, and tariffs create barriers to investment and can deter stakeholders from committing to large-scale solar projects. Additionally, trade disputes and geopolitical tensions can disrupt supply chains, leading to fluctuations in material costs and constraining the availability of key components such as silicon wafers and photovoltaic cells [33]. Technological hurdles also abound, with research focusing on enhancing solar cell efficiency, durability, and scalability while simultaneously driving down production costs to compete with conventional energy sources [34]. Material innovation is crucial, particularly in reducing reliance on rare earth elements and toxic substances while improving recyclability and sustainability across the solar value chain [35]. Grid integration poses another significant challenge, as intermittent solar energy generation requires robust infrastructure upgrades and advanced energy storage solutions to ensure reliability and stability within existing power grids [36].

Furthermore, societal perceptions and public acceptance play pivotal roles, with concerns over land use, visual impact, and environmental footprint shaping the social license for solar projects [37]. Addressing these challenges necessitates a holistic approach, involving collaboration among governments, industry players, research institutions, and communities to foster a conducive environment for sustainable solar energy growth. International cooperation through initiatives like technology sharing, knowledge transfer, and capacity building can accelerate progress and facilitate the global energy transition towards a cleaner, more resilient future powered by solar energy.

6.3 SOLAR CELL MARKET

The dynamics of markets, particularly in the context of emerging technologies such as photovoltaic (PV) systems within the energy sector, are deeply influenced by a complex interplay of economic principles, institutional frameworks, and governmental interventions. Traditional market theory posits an ideal scenario where markets operate efficiently, driven by the forces of supply and demand, with minimal government interference. However, the reality often diverges from this ideal, as markets are subject to various barriers and influenced by institutional factors that shape their development. The global solar photovoltaic (PV) industry has witnessed significant growth over the past few years. In 2023, it is estimated that approximately 330 gigawatts (GW) of solar PV installations will be added worldwide, according to the Solar Energy Industries Association (SEIA) [38]. This marks a substantial increase from the previous year, with 262 GW installed in 2022, as reported by the International Energy Agency (IEA). Despite the challenges posed by the COVID-19 pandemic and supply chain disruptions, the solar PV market continued to expand [39]. Additionally, the average installed price of solar PV systems has been decreasing over the years.

In 2023, estimates from the National Renewable Energy Laboratory (NREL) suggest that the average installed price ranged from \$2.8 to \$3.2 per watt, showcasing a downward trend from \$3.0 to \$3.5 per watt in 2022 and \$3.3 to \$3.8 per watt in 2021 [40]. This decline in prices is attributed to technological advancements, increased competition, and supportive policies. The top solar markets, including China, the USA, and India, have consistently dominated global installations, with China leading the way in capacity additions. Despite variations in annual growth rates, these countries remain key drivers of solar PV deployment, contributing significantly to the global transition towards renewable energy [41]. In examining the barriers to entry faced by firms venturing into the energy market, it becomes apparent that these obstacles are multifaceted and can stem from financial limitations, market conditions, regulatory complexities, and firm-level challenges [42]. Moreover, the presence of high entry barriers may lead to market structures characterized by oligopolies or monopolies, restricting competition and innovation.

Economic sociology sheds light on the role of institutions in shaping market dynamics, emphasizing that markets are not always efficient due to the influence of formal regulations and informal norms. Institutions, which encompass human-made constraints governing economic interactions, play a crucial role in determining market behavior and structure [43]. Governments, as architects of formal institutions, wield significant power in regulating and shaping market structures. However, their interventions can introduce instability and uncertainty, potentially disrupting market functioning. Government interventions in the market can be guided by both economic considerations and political motives. Economic interventions aim to address market failures and promote efficiency, while politically motivated interventions seek to align market functioning with broader societal goals. Such interventions, whether in the form of taxes, subsidies, or regulatory measures, inevitably impact market participants and can influence the trajectory of market development. The adoption of technological innovations, such as PV systems, holds immense potential for transitioning toward renewable energy production and consumption [44].

6.4 SOLAR CELL APPLICATIONS

6.4.1 Green Hydrogen Solar Plant

Hydrogen is considered a clean, pollution-free, and sustainable energy source. Its emission gives only water as a byproduct and no carbon-based harmful compounds. A green hydrogen solar plant uses solar energy to produce hydrogen. This method is considered "green" because it relies on solar energy to generate electricity, which is required for the splitting of water molecules into hydrogen and oxygen via electrolysis [45]. The first solar energy-based green hydrogen production was established in Madrid, Spain in 2000. Nowadays, many countries like the U.S., Japan, Canada, Iceland, etc. have incorporated green hydrogen plants to improve the efficiency of their energy systems [45]. Green hydrogen produced herein can be used in a variety of applications like transportation, oil production, petrochemical refining, ammonia production, fuel cells to generate electricity, etc.

A typical setup of a green hydrogen solar plant requires solar panels, an accumulator battery set, an electrolyzer, and hydrogen storage containers, as shown in Figure 6.1.

The solar energy panels or cells capture sunlight and convert it into electricity. Generally, silicon-based cells are used for this process whose efficiency ranges from 12% to 15%. However, for GaAs (Gallium Arsenide), solar cells have high efficiency of 25%–30%, and the total efficiency for the whole process is approximately 16% [46]. Sometimes a battery setup is attached to store solar electricity. This stored electricity can be used further in the electrolysis step. Three types of technologies are used to conduct water electrolysis: the alkaline process, the proton exchange membrane (PEM) process, and the solid oxide process [47]. In an alkaline process, KOH, NaOH, or NaCl are used as electrolytes. The temperature range in this process is 20°C–80°C with 59%–70% efficiency. The electrodes are made up of metals like nickel with an electrocatalytic coating. It requires a gas-impermeable separator to prevent the product gases from mixing. Hydroxyl ions (OH^-) are transported to the anode, where they are converted into water and oxygen. Reduction of water takes place at the cathode to produce hydroxyl ions and hydrogen. This process has a lot of advantages like low cost, higher lifetime, and mature technology. The disadvantages include limited current densities, crossover of gases, and high concentration of KOH solution.

In the PEM process, a polymer such as Nafion and water are used as electrolytes. In Nafion, the sulphonic acid groups are attached to the tetrafluoroethylene backbone. Upon hydration, these groups act as proton donors. Noble metals like platinum and iridium oxide act as the electrocatalysts in this process. At the anode, water splits into oxygen, H^+ ions, and electrons. H^+ ions are then transported to the cathode via sulphonic acid groups of PEM. At the cathode, the reduction of H^+ ions takes place to give hydrogen gas. The temperature range for this process is 20°C–80°C with 70% efficiency, similar to an alkaline process. The major disadvantages of this process include the high cost of the components, i.e., polymer, electrode materials, and current collectors. But this process is safer as compared to the alkaline process.

In the solid oxide process, yttria-stabilized zirconia is used as an electrolyte. The cathode is made up of yttria-stabilized zirconia and Ni. The anode is made up of perovskite materials. At the anode, the oxide ions (O^{2-}) are oxidized to generate oxygen and electrons. At the cathode, the reduction of water takes place to form hydrogen

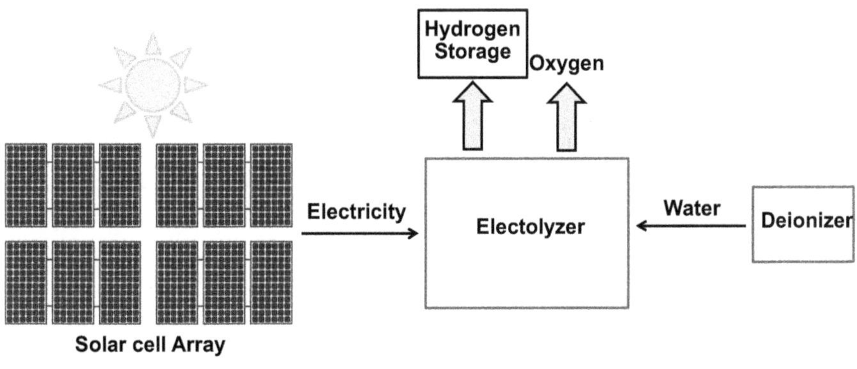

FIGURE 6.1 The process of formation of green hydrogen.

and oxide ions. This electrolyzer operates at very high temperatures (500°C–850°C) with water in the form of steam with almost 100% conversion efficiency. The advantages of this process include high efficiency and non-requirement of noble metal electrodes. Despite 100% conversion, this process has short-term stability.

Water used in all these methods must be purified and should not contain any minerals. Because when saline water is used, then chlorine gas is produced instead of oxygen. Hydrogen produced by electrolysis is collected and stored in a container. Oxygen produced can be stored in containers or can be released back into the air.

6.4.2 Space-based Solar Power Satellites (SPS)

A space-based solar power satellite (SPS) is a concept proposed in 1968. It would use solar cells placed atop satellites pointed toward the sun to generate power while rotating the Earth in a geosynchronous orbit. The electricity generated is subsequently transformed into microwaves, which are sent to the Earth's surface via Earth-pointed wireless power transmitter (WPT) technology. Its development intends to address future global environmental concerns, such as lowering greenhouse gas emissions, by providing a more sustainable alternative to traditional energy sources. It can harvest solar energy regardless of weather conditions, regional boundaries, or the day-night cycle and delivers better power densities and higher power output per unit area over the ground-based systems. It might potentially meet security and military objectives by providing steady and reliable power sources for essential infrastructure such as communication networks, military outposts, and remote areas. During operation, it emits no pollutants or greenhouse gases, making a substantial contribution to environmental conservation [48].

An object revolving around the Earth is named as a satellite. The man-made satellite is referred to as an artificial satellite. Such satellites are used for various purposes such as communication, navigation, weather forecasting, earth observation, research, military, and security. Classification of these satellites is based on the objective and orbit on which they revolve [48]:

a. Low Earth orbit (200–2000 km)
b. Medium Earth orbit (2000–35,785 km)
c. Geosynchronous orbit (35,786 km)
d. High Earth orbit (above 35,786 km)

Among all these orbits, the geosynchronous orbit is considered one of the best for collecting solar energy, as it is capable of capturing 96% of the sun's rays. The International Academy of Astronautics (IAA) examined three types of SPS architectures: SPS Type-I, SPS Type-II, and SPS Type-III.

a. Type-I: SPS Type-I consists of one or more sun-pointed solar energy collecting systems and an earth-pointing wireless power transmission (WPS) that uses microwave RF for power transmission. The major issues are high-voltage power management and distribution systems and initial infrastructure is required both in space and earth-to-orbit (ETO) transport.

b. Type-II: In this type of SPS, the microwave technology used in SPS Type I is replaced by laser technology, which involves electric laser-based and solar-pumped laser systems. The concept of electric laser-based is feasible. In this approach, the system comprises solar cells for power generation with power transmission by lasers and specially designed photovoltaic systems. The major concern is the end-to-end efficiency which is lower than WPT.
c. Type-III: This design, known as the SPS sandwich, employs a highly modular architecture, which uses a light re-directional approach for the distribution of energy, large-scale solar power, PMAD (power management and distribution), and WPT systems. In this approach, microwave RF is received by large ground-based rectenna systems and addresses the PMAD issues encountered in the concept of Type-I. This concept is feasible with currently available technology, achieving efficiencies slightly above 50%. Further technological improvements are necessary to enhance both the efficiency and cost-effectiveness of wireless technology [49].

Currently, three modes of wireless power transmission are based on microwave, laser, and hybrid. In the case of microwave-based transmission, solar cells are mounted on the satellites to generate electricity directly from solar light. This electricity is then converted into microwaves using magnetrons for transmitting the energy to the ground at a microwave frequency of 5.6 MHz. In this approach, these microwaves are collected on the ground by rectenna in the form of DC, which is then converted into AC (alternating current) for domestic and commercial purposes.

In the concept of lasers, the energy harvested from the sun is converted into monochromatic light using laser technology. This monochromatic light is received on the ground by photovoltaic solar cells, which convert the light into electricity. This technique theoretically reduces efficiency losses observed in the generation of solar power.

The hybrid technique integrates the benefits of both laser and microwave technologies, minimizing the drawbacks of each technology and maximizing their advantages. Hybrid ideas would make the best possible system. When environmental attenuation is limited, a laser approach is recommended. Subsequently, the transmission transitions to microwave radio frequency (RF), which is considerably less affected by atmospheric attenuation.

A rectifying antenna, referred to as a rectenna, is a specially designed receiver to harvest wireless power through microwave radio frequency (RF) by converting it into direct current (DC) [50]. The block diagram of receiving RF and its conversion to DC is illustrated in Figure 6.2.

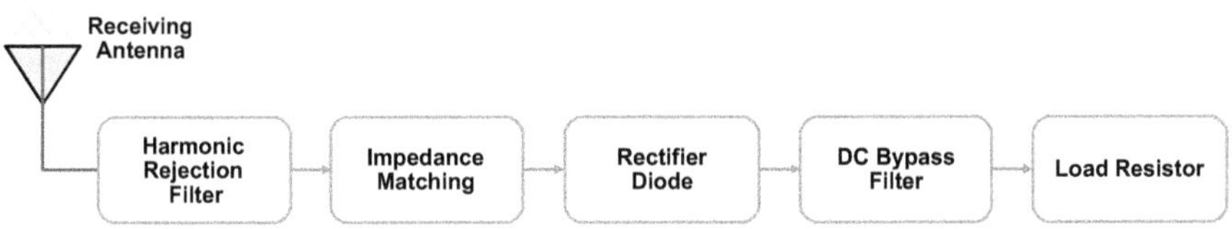

FIGURE 6.2 Schematic of rectenna.

The harmonic rejection filter prevents energy from dispersing into free space, which could create interference with surrounding electronic devices operating in the same frequency band as the rectenna. The impedances of the transmitter and receiver must match for optimal power transfer. Impedance matching circuits match the impedance of the receiver to that of the transmitter to ensure compatibility. RF signals are converted into DC electricity using rectifier diodes. This DC electricity is not pure since it includes an AC component. Filtration must be carried out to provide pure DC, which can then be transformed into an AC signal as desired.

The most optimal frequency range for power transmission in space via microwave transmission is from 1 to 6 GHz. The efficiency of power transmission decreases beyond this range due to increased attenuation. Often, either 2.4 or 5.8 GHz frequencies are utilized and are most effective for the transmission of electricity via microwave RF for Industrial, Scientific, and Medical (ISM) purposes. On the other hand, the laser technique uses a wavelength of about 1.06 μm for power transmission from space to the ground [51]. This particular wavelength is capable of sending power over large distances and can also pass through the Earth's atmosphere.

Before commencing any project, it is critical to evaluate capital expenses. This is especially true for space-based solar power systems, where all costs, including launch, maintenance, and operations, must be carefully planned and accounted for. Economic feasibility is critical, as unreasonably high unit costs would render the technology inaccessible to the general population. The primary hurdle is the launch cost, which includes transferring multiple components required for assembling the plant in space. Many countries do not undertake such expensive projects due to the hefty expense. Researchers are stepping up efforts to find innovative approaches that help reduce expenses and increase the economic feasibility of space-based solar power generating systems.

6.4.3 Solar Cells in Industry

Electricity is the backbone of industry, and development within industries is not possible without it. A major fraction of the world's energy is utilized by industries. Manufacturing, agriculture, construction, and mining are the dominant sectors in energy usage. The usage of energy is increasing year by year due to industrial development [52]. In the present scenario, where fuel prices are rising and restrictions from environmental departments, industries are trying to stay away from carbon and greenhouse emission fuels. There is a need to switch from fuel to renewable sources with the development of new technological development. Various renewable sources are available, such as solar, wind, and hydro. However, solar energy has garnered more attention in comparison to other renewable energy sources due to its abundance, relatively low installation costs, economical maintenance, noise-free functioning, and easy installation. Various techniques are applied to harness solar energy. Generating electricity from solar energy using photovoltaic cells in industries is very common. In textile industries, two major processes are dyeing and drying in which hot water is required. Solar energy is a better solution and a step towards a sustainable future [53]. In food industries, a variety of processes, such as drying and dehydration processes, preheating, concentration, pasteurization, and sterilization, are involved in preparing, preserving, and maintaining the nutrient quality of food. In all these processes, solar energy can be utilized to reduce carbon footprints [54]. Indeed, solar energy is versatile and extends its applications beyond basic heating and generating electricity in various industries. These applications include washing, cleaning, chemical reactions, industrial space heating, building, desalination, cooling, enhanced oil recovery, hydrogen production, and solar ponds from solar thermal energy.

Utilization of solar energy as thermal heating and cooling:

a. Thermal heating: In this terminology, solar energy is collected in the form of thermal energy. This energy is utilized for a variety of tasks, such as providing water heating, building heating, and pool heating for residential, commercial, and industrial applications. By the end of 2022, estimated global solar thermal capacity reached 542 GWth which was 62 GWth in 2000. Figure 6.3 represents the basic schematic diagram for the conversion of solar energy into mechanical energy. This system consists of solar collectors and concentrators to gather solar radiation, and the stored energy is used for domestic, commercial, and industrial units.
b. Solar thermal cooling: In this technology, heat collected from the sunlight is harnessed to power the refrigeration and other cooling systems for domestic and commercial purposes through thermally driven cooling procedures. This procedure relies on the principles of thermodynamics to extract heat and lower the temperature without the use of conventional electricity. A block diagram featuring its primary components is presented in Figure 6.4. It consists of three major components: solar collector, cooling system, and heat rejection system. Solar collector captures the solar energy and convert it into heat or another form, which is routed to a cooling system, such as refrigeration, and finally, this heat is rejected to sink.

In general, the following three techniques are preferred for harnessing solar energy for water heating and producing steam in industries as required:

a. Parabolic trough collector (PTC): In this method, solar radiations are focused onto the receiver tube using concave mirrors. The focal line, where the receiver pipe is positioned, receives concentrated sunlight and raises the temperature of the fluid within the receiver pipe. The operating temperature

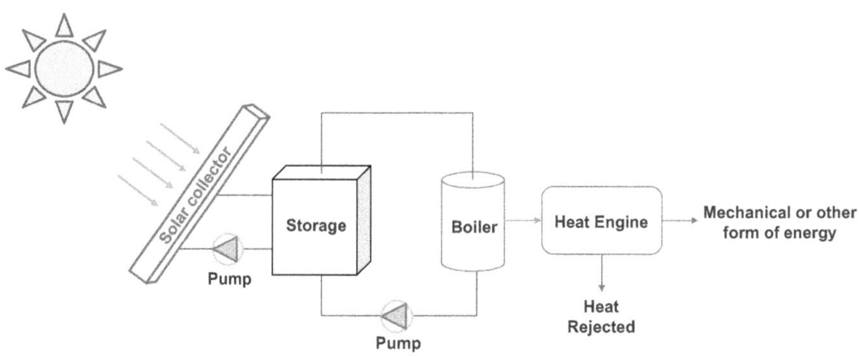

FIGURE 6.3 A schematic representation of a solar thermal conversion system [55].

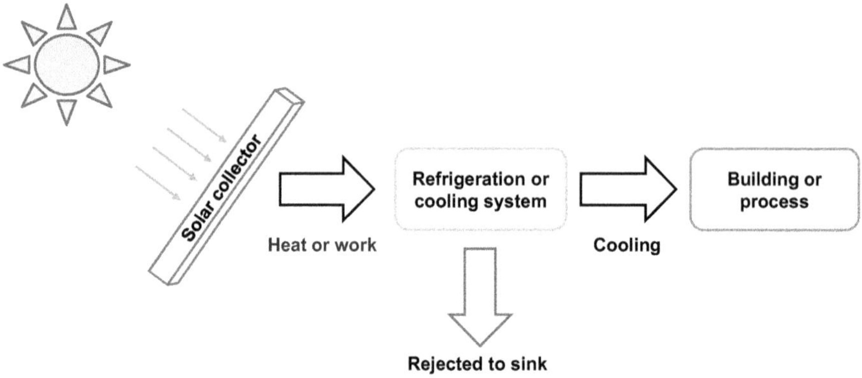

FIGURE 6.4 Process of flow in a solar cooling system [56].

of the system exceeds 500 K. This technology is cost-effective but has less accuracy. It finds applications in various industries for harnessing heat from solar energy [57]. A schematic of a parabolic solar concentrator is represented in Figure 6.5.

b. Linear Fresnel reflector (LFR): The technology is named after the Fresnel lens. This technology consists of linear reflecting arrays which are arranged to focus solar radiation onto a focal line. Usually, flat-plane mirrors are used to track the sunlight and focus on an absorber tube [58]. An additional compound parabolic reflector is also used to direct the sunlight onto the absorber tube. A basic model of LFR is shown in Figure 6.6. Various other

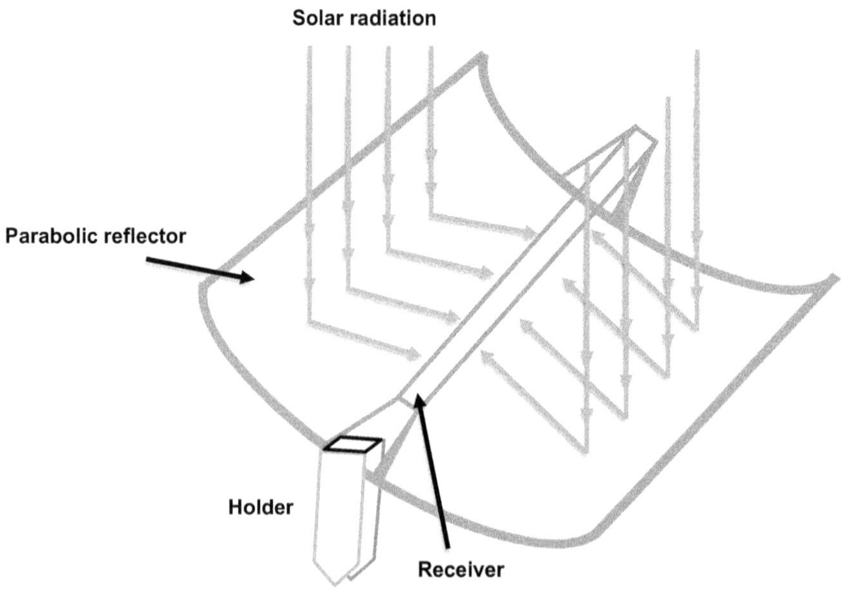

FIGURE 6.5 A schematic of the parabolic trough collector.

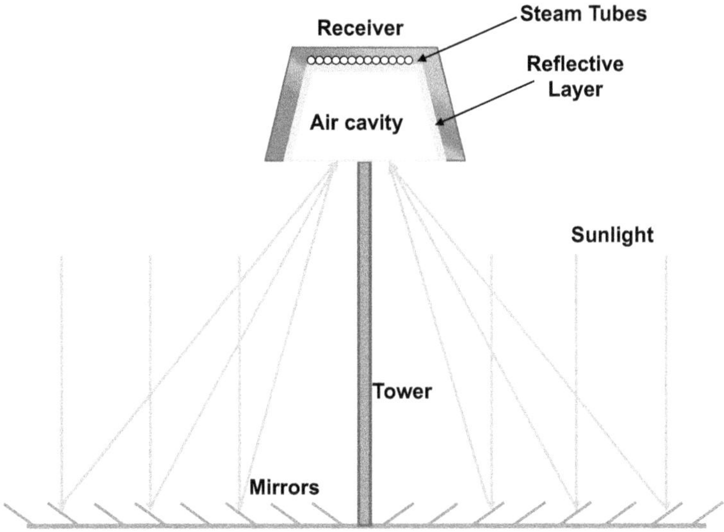

FIGURE 6.6 LFR with a central tower as receiver [58].

types and models are also available depending upon the requirements and other parameters for which they are used.

c. Evacuated solar collector: This is the most efficient and sustainable method by which solar energy is harnessed for heating water. The collector consists of multiple glass tubes. Each tube consists of a small glass tube. A vacuum is created between the gap or space between the outer and inner glass tube. This vacuum acts as thermal insulation and prevents heat loss absorbed by the absorber plate. The inner surface of the tube is coated with a material that absorbs more heat. This tube contains a special fluid, referred to as heat transfer fluid (HTF). As the solar radiation heats up the absorber tube, the fluid evaporates and rises to a condenser section. This heat is then transferred to the water flowing through the manifold. After losing its heat, the condensed fluid returns to the lower section of the pipe as shown in Figure 6.7. This repeated process heats up the water [59].

Industries employing the aforementioned techniques:

1. Solar desalination: Due to advancements in industrialization, modernization, and life standards, the demand for freshwater is increasing proportionally. Fresh water is unable to meet our demands due to limited reservoirs. The United Nations Organization reported that by 2025, 1800 million people will be affected by the clear water crisis. To overcome this situation, desalination technology is employed. However, this technology relies on fossil fuels and has adverse effects on the environment by emitting greenhouse gases. Renewable energy-powered technology can mitigate the adverse effects up to certain limits.

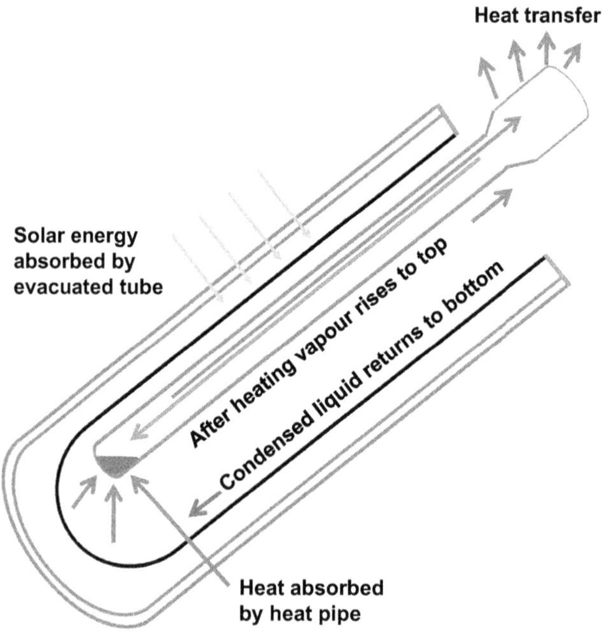

FIGURE 6.7 A simplified diagram of an evacuated tube solar collector [59].

A basic solar desalination system comprises an inclined transparent sheet. This sheet facilitates the transmission of solar radiation and the condensation process over a water basin. In order to absorb more radiation from the sun, this basin is coated with black color. It results in heating up water for evaporation, and after condensation, vapors are collected along the inclined glass. Various methods are employed to increase the condensate yield, including coupling various collectors such as flat plate collectors, evacuated tube collectors, and concentrating dish-type collectors. In addition to these, phase change materials and photocatalyst coating also enhance condensate yield. The material of the cover and angle of inclination, as well as the coating material of the basin, also impact the distillation process. Various models of solar still are available [60–62]. Figure 6.8 illustrates the schematic representation of a basic solar still.

2. Textile industries: The textile sector primarily focuses on creating and designing yarn, fabric, clothes, and, ultimately, distributing the final products in the market. The market of textile industries in the world is about 4000 billion US dollars [63]. To produce the final products, it undergoes various processes, including, spinning, weaving, and finishing, which are further subdivided into various procedures, as illustrated in Figure 6.9 in the form of a flowchart. The energy utilized during these processes is significant and mostly depends on fossil fuels. The required temperatures for various processes range from 40°C to 120°C,

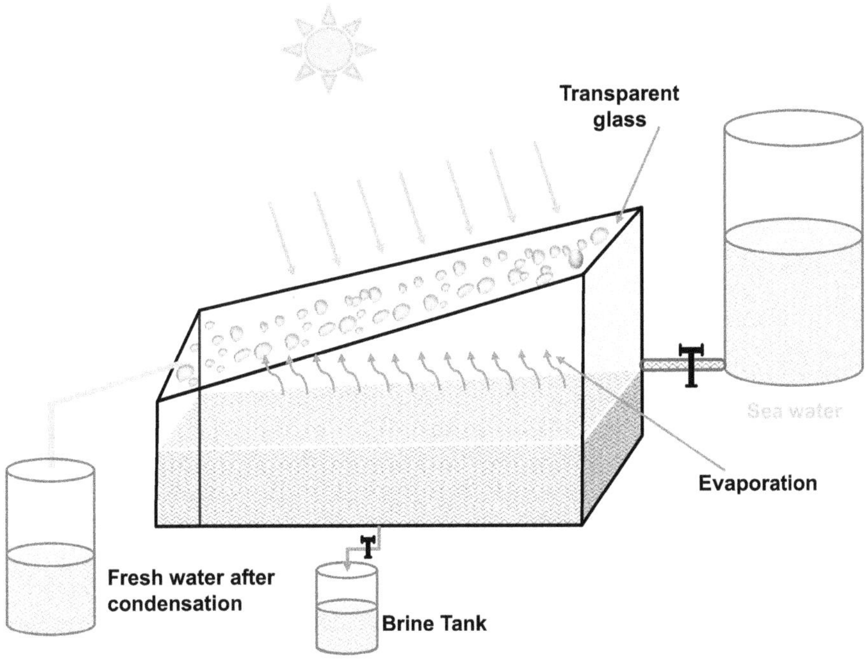

FIGURE 6.8 A schematic of basic solar still [62].

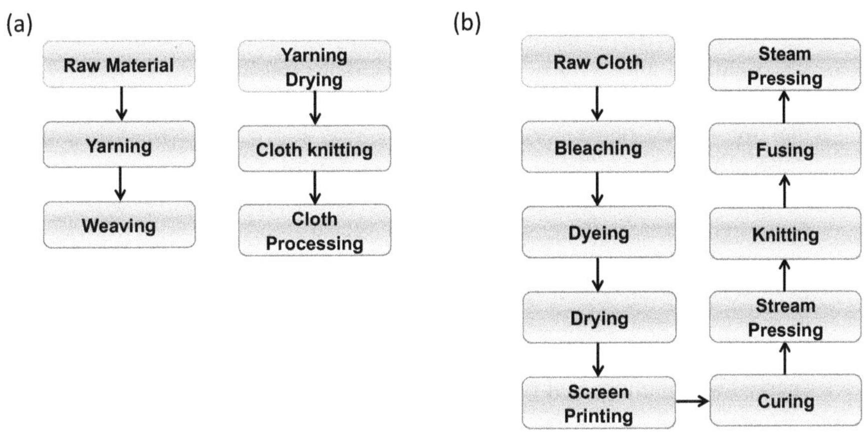

FIGURE 6.9 Flow chart of (a) textile manufacturing and (b) Garment manufacturing [63].

as presented in Table 6.1 [64]. Mainly, steam and hot water are used for various processes. Solar energy has the potential to replace traditional fuels and transform this industry into a more sustainable and ecologically friendly one. To attain a temperature around 120°C, an evacuated tube collector is preferred. For higher temperatures, a linear Fresnel reflector is the optimal choice (Table 6.1).

TABLE 6.1
Processes and Temperature Requirements [64]

Process	Temperature Required (°C)
Bleaching and dyeing	60–90
Drying and degreasing	100–130
Dyeing	70–90
Fixing	160–180
Pressing	80–100

3. Paper industries: Pulp is the raw material for manufacturing papers. A pulp mill produces a slurry made from various materials, including rags, cellulose fiber from wood, wood chips, waste paper, straw, and hibiscus bark. After pulping, it undergoes bleaching, then chemical recovery, and finally paper making. For various processes, heat and steam play a significant role and the temperatures required for these operations vary from 125°C to 200°C. Instead of using fossil fuels and other non-renewable sources to attain this temperature, renewable sources, such as solar energy, can be integrated to make this industry less pollutant. Such an intensive power industry uses captive power, where the power generated by the industry is consumed by itself, using solar techniques and harnessing heat directly from solar energy using ETC, LFR, and PTC [63,65]. The fluctuations in solar radiation are reduced by thermal buffer storage.
4. Food industry: The food industry is primarily concerned with the processing of edibles and drinks. It includes industries related to dairy products, beverages, poultry, meat, and seafood processing, confectionery, candy making, spices, pharmaceuticals, and canned and frozen food. Chilling, sterilization, and spraying are basic processes in the food industry. Hot water and steam are utilized in major food industries. ETC and LFR can fulfill the necessary heat requirements of various food industries [63,66].
5. Institutional solar cooking: Normal kitchens are designed for family use. However, cooking for a large group of people in the same kitchen is referred to as institutional cooking. Institutional cooking involves preparing meals for hospitals, hostels, orphanages, hotels, temples, and other community centers, where traditional fuels such as kerosene, LPG, and wood are often used, releasing significant amounts of carbon and greenhouse gases [66]. To combat this issue, one of the greatest ways is to cook with non-polluting renewable sources, particularly solar energy. As we are aware converting one form of energy into another usually results in loss to a significant level. High-concentrating solar energy can be used for cooking. The contracting parabolic mirrors are used to focus sunlight to attain high temperatures ideal for cooking. Such a solar-steam cooking system based on "Scheffler dishes" is installed at Shri Saibaba Sansthan Trust, Shirdi, Maharashtra, and Brahma Kumaris, Mount Abu, Rajasthan, India. The average daily number of pilgrims visiting the Saibaba Temple reaches 50,000, with an annual average of approximately 15 million visitors from various parts of

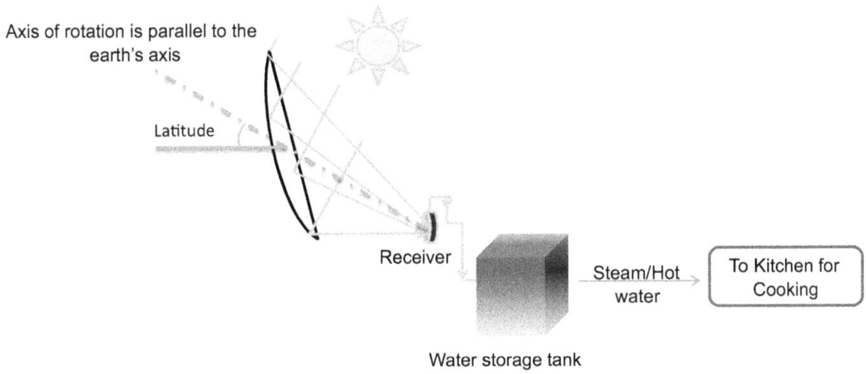

FIGURE 6.10 Schematic of Scheffler dish type solar thermal cooking system [66].

India and abroad. Brahma Kumaris claims 0.3 million visitors annually. This solar cooking system is efficient in raising the temperature of the water and then converting it into steam for cooking purposes. It consists of parabolic reflectors, known as Scheffler reflectors, which are used to focus on the receiver. Solar energy is absorbed by the receiver and transfers heat to the water flowing through it. The water is then converted into steam and stored in steam storage tanks. This steam reaches the kitchen through thermally insulated pipes for cooking. Multiple Scheffler reflectors are required depending on the demand for cooking [67–70]. A schematic of Scheffler's steam-based cooking is demonstrated in Figure 6.10.

6.4.4 Solar Panels All over the Rooftops in Family Housing

Solar collectors and panels are most often installed on the rooftops of buildings/households, on their facades, or on the ground in many countries as a sustainable and affordable alternative for gaining access to electricity. In fact, rooftop solar systems are contributing to the electricity generation of many countries. For example, the European Union's rooftop solar system contributes 64%, the U.S. 31.4%, China 27%, and Australia 25.8% of its solar generation [71].

The electricity demands in the buildings/households mainly rely on the grid electricity and hence a large amount of electricity bill to pay. So, the installation of solar PV systems for individual households is becoming more popular in many countries [71–73]. Hence, it is expected that the installation of solar panels in buildings will increase in the future due to decreased capital costs and stronger governmental energy policies [72]. Researchers found that types of motivation, economic benefits, demographic characteristics, government energy policies, and financial support affect the adoption of residential solar PVs by low-income families, as well as local social housing contextual considerations [74,75].

The complete rooftop setup for solar electricity production includes solar panels, inverter, and monitoring system, as shown in Figure 6.11. The monitoring system is highly important because it helps in checking systems performance and efficiency. The solar panels are made up of silicon mostly. They convert solar energy into direct

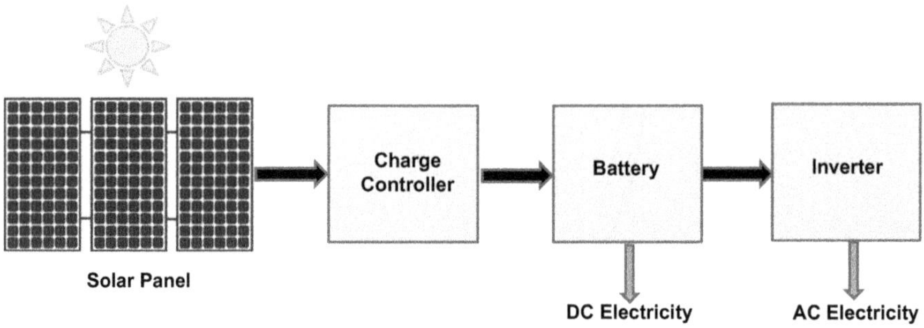

FIGURE 6.11 Block diagram of rooftop solar panel system.

current (DC). The inverter converts this DC into AC. This AC electricity is then used in the house or fed back to the grid. Solar batteries store surplus energy generated by solar panels if not fed to the grid. This energy can be used during the night or in situations when solar energy is not available.

Three types of batteries can be used: (a) Lead acid batteries (particularly Deep Cycle batteries), (b) Lead carbon batteries, and (c) Lithium batteries. Out of these three, deep-cycle batteries are used mostly due to their cost-effectiveness and reliability. Deep cycle batteries are designed to withstand repeated charging and discharging cycles, making them well-suited for solar power systems. These batteries can be discharged upto 50%. This ensures longevity and optimal performance over multiple charge-discharge cycles. These batteries contain a metal plate made of lead and antimony (as anode), a metal plate made of lead dioxide (as the cathode), and a sulfuric acid-water mixture (as an electrolyte). Deep cycle batteries contain plastic cases and hence can be stacked in series. One battery generates 2 V and if 12 V is required, six batteries can be wired in series [76,77]. The charge controller manages the connection between the solar panels and the batteries by acting as a switching device. The charge controller ensures that charging occurs at the correct voltage and protects the batteries from overcharging by automatically cutting off charging when the batteries reach full capacity. This proactive approach helps prevent overheating and damage to the batteries' internal components [78].

There are three types of solar systems for rooftop installation. (a) On-grid systems: This system is integrated with the main grid supply. The main grid supply is used only when the rooftop solar system is not able to supply the required power. So, in this way, a well-planned rooftop solar system can save a lot of electricity bills. If power is generated in excess, then it can be put into the grid, for which compensation can be obtained from the government. (b) Off-grid systems: This system is not integrated with the main grid supply. Power generated from solar systems is stored in a battery, which is used when there is no grid supply. (c) Hybrid systems: in these systems, both on-grid and off-grid systems work. In this type of system, a battery is used. The advantage here is that after the battery is fully charged, the excess power generated is fed to the grid which generates additional revenues for the consumer.

For a small to medium-size household, a minimum of 100–150 square feet of rooftop area and solar panel of 3–5 kW is sufficient. A 3–5 kW system can produce 4000–6000 kilowatt-hours (kWh) of electricity per year. The high initial cost and proper maintenance are the main drawbacks of this system. This is the main reason that adoption of rooftop solar has been concentrated in higher-income households only. However, as the cost of solar systems is declining and the market is expanding, the focus is shifting towards increasing solar access to low-to-moderate income households [79].

6.4.5 Rolling Solar Panels

Rolling solar panels comprise solar cells that can easily roll or bend. These solar panels roll up to form a compact cylinder just like a tape. One such example of the silicon-based rolling solar panel is shown in Figure 6.12, which was made by Liu et al. [80]. These solar panels have several advantages over conventional solar cells. Conventional solar cells possess very high thicknesses. Hence, these cells are not suitable for some applications like satellites, spacecraft, and unmanned aerial vehicles. For these applications, very thin and flexible solar cells are required. These flexible solar panels have also been incorporated in many automobiles. For example, solar panels have been put on the engine compartment hood, passenger cabin roof, and cargo cabin door in Toyota Prius [81]. As these solar panels can easily roll, so their transportation is very easy. Hence, these solar panels can be used as power sources for remote military operations, remote locations, and natural disasters. It is expected that these flexible solar panels will be put in many mobile and electronic device chargers to charge various electronic equipment like electric cars, electric bicycles, smartphones, and laptops in the near future.

These solar cells are fabricated on flexible substrates such as metal, ceramic, and plastic/polymer [82]. A flexible substrate possesses three characteristics: (a) it bends or rolls without plastic deformation, (b) its shape does not change under various conditions, and (c) it is able to stretch elastically. Thin metal foils such as stainless-steel

FIGURE 6.12 Photograph of a 15.6 cm × 15.6 cm flexible silicon solar cell.

foil are most commonly used due to their low cost and thermal & chemical stability. Its use started in the 1980s and is used till now as well [83]. Titanium foil has also been used in some solar cells. But due to its high cost, it cannot be used in commercialised solar cells [84,85]. In the case of ceramics, glass is the most commonly used substrate due to its stability in heat, chemicals, and moisture. Flexible glass substrate of less than 100 μm thickness has been used by various researchers [86,87]. Plastic/polymer substrates, due to their lightweight and low cost, have been used in many solar cells [88,89]. However, the high permeability of plastic toward moisture and oxygen can destroy solar cells. So multilayer coating is required. Another drawback is the low thermal stability of polymers. The photovoltaic material layer used in these solar panels is made up of either bulk heterojunction or bilayer devices (Figure 6.13) containing silicon, CIGS, organic polymers such as poly(3-hexylthiophene-2,5-diyl) (P3HT) as electron donor and fullerene derivatives as electron acceptor layers, or single layer perovskites (Figure 6.13c), etc. Two kinds of electrodes are used in these panels. One electrode is made of transparent conducting oxides such as ITO (indium tin oxide). This electrode is made transparent so that light can enter the solar cell. Another electrode is made up of ultrathin metal films such as Al, Ag, or Mg. Since the thickness of these metal electrodes is less than 20 nm, these are also semitransparent. In some cases, both the electrodes are made transparent by using ITO. In the case of inorganic solar cells, metal grids are used as front layers. All these layers are assembled on a flexible substrate with the help of techniques like vacuum-phase deposition, solution-phase spin-coating, and printing [90].

When these solar panels are exposed to sunlight, the photovoltaic material absorbs photons, and excitons are created. These excitons are then transported to the donor-acceptor interface, where excitons split into free electrons and holes. Free

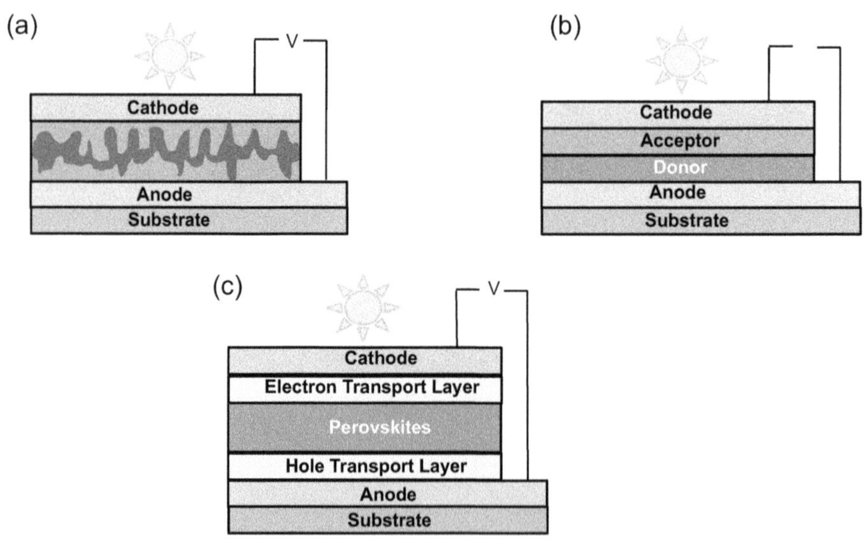

FIGURE 6.13 Working of rolling solar cells as (a) bulk heterojunction organic, (b) bilayer organic devices, and (c) perovskite based device.

electrons are then transported to the n-type layer, and holes are transported to the p-type layer. The difference in charge carrier concentration creates a current known as photocurrent. This current can be used directly or can be into electrical systems namely batteries or an inverter for storage. As these panels are rollable, they can be easily installed or retracted.

Although these panels have a lot of advantages like their portability, flexibility, lightweight, ease of installation etc. But the disadvantages like their low efficiency, low durability, and high cost need to be addressed.

6.4.6 Transparent Solar Cell

Transparent or see-through solar cells are the latest type of solar cells, which are clear like glass. Transparent solar cells allow a significant portion of light to pass through them, enabling their integration into various surfaces like windows, screens, or even smartphone displays. Hence, the most important use of this type of solar panel is in Building Integrated Photovoltaics [91]. If the building containing glass on its surface replaces 90% of its glass with transparent solar panels printed on glass, then 40% of the building's energy requirement will be fulfilled by these panels. Transparent solar cells can be incorporated in the glass surface of automobiles so that they can be powered with solar energy. These can also be used in many electronic devices.

These solar cells use glass as a substrate. A very thin layer (less than 20nm) of ITO is deposited on the glass. The coating of a very thin layer makes the solar cells transparent. After this, a very thin film of photovoltaic material is deposited. This deposition can be done by many techniques, such as chemical bath deposition, physical vapor deposition, electrodeposition, screen printing, pulsed laser deposition, and spray and atomic layer deposition. Out of all these methods, screen printing is considered as best method as transparency can be controlled by this method. Screen printing is a technology used for the formation of front and rear metallic contacts in crystalline silicon solar cells. But nowadays, screen printing can be used to prepare all the functional layers [92–94]. In transparent dye-sensitized solar cells, the TiO_2 layer is deposited by screen printing. The efficiency of this cell is only 9.2%. The main problem with transparency is that most of the photons get passed through and not absorbed. This problem was taken into account by Lunt and Bulovic [95] who prepared organic solar cells that can absorb ultraviolet and near-infrared radiations (650–850nm) and visible light is passed through. But their efficiency is only 1.3.

In some other reports, the transparent polymer has been combined with transparent silver nanowires to create transparent solar cells, as shown in Figure 6.14 [96,97]. The resulting solar cell also absorbs ultraviolet and near-infrared radiation. Transparent fluorescent dyes which absorb ultraviolet and near-infrared radiations and convert them into visible radiations, have also been used. Semi-transparent solar cells have also been prepared which consist of perovskites as photovoltaic material [98,99].

When light strikes the transparent solar cell, UV and NIR light is absorbed by the cell, and visible light is transmitted. The reflector sends UV and NIR light back into the active layers, while the anti-reflective (AR) coatings on the outside surfaces maximize incoming light by reducing reflections. The photons with enough energy can excite electrons within the material, creating an electron-hole pair. The separated

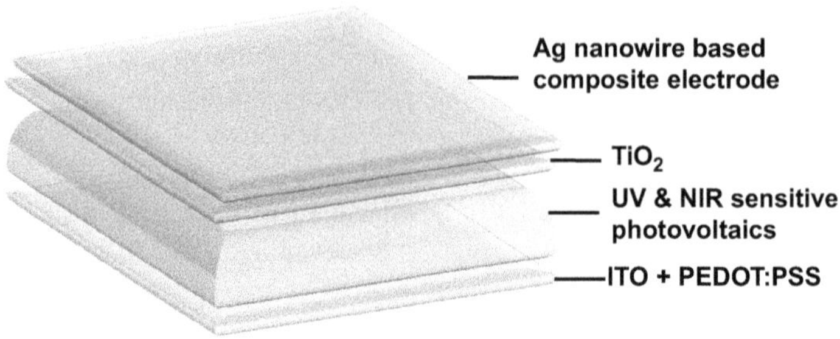

FIGURE 6.14 Composition of transparent solar cell.

electrons flow through a conductive pathway, creating an electrical current. This current can be used to power electronic devices or stored in batteries.

The lesser efficiency and high cost of these solar cells limit their applicability. Researchers are actively working on improving the efficiency of transparent solar technology.

6.4.7 Solar Cells in Calculators

Solar-powered calculators are a sustainable alternative to battery-powered devices. These calculators were first introduced by Sharp in 1976. The major disadvantage of these devices was the requirement of a good amount of light to function properly. So, at that time it was difficult to use these calculators in poorly lit areas or during night-time. This problem was resolved by Texas Instruments in the 1980s using Anylite Technology. They made calculators by the incorporation of large solar cells to solve the problem. The Anylite technology in Texas instruments solar powered calculators is indicated by the lowercase alphabet 'a' in modern devices. Older devices have Anylite Solar impressed on them [100].

The primary components of solar calculators include a liquid crystal display, a small solar cell (generally made of amorphous silicon with a voltage output from 1.5 to 2 V) located above the display screen, and a small rechargeable battery for use in low-light conditions. Solar cell captures light (both from artificial and natural sources) and converts this light into electrical energy through the photovoltaic process. This electrical energy is either used immediately to power the calculator or stored in the battery. As these calculators need very little power, so even indoor lighting is sufficient to keep them operational. If the battery gets discharged, there occurs a time gap between pressing the number and the change in display. These calculators must be protected from direct sunlight as direct sunlight can damage the device due to heat absorption and overvoltage generation. Solar panels in these calculators must be regularly cleaned, and oil, and dirt, if accumulated, must be removed for their proper functioning.

6.5 CURRENT CHALLENGES AND FUTURE SCOPE OF SOLAR CELL

Solar cells are rapidly evolving, driven by technological advancements, increasing demand for renewable energy, and concerns about climate change. Solar cell technology has grown enormously in recent years, leading to both current challenges and future scope. Recent breakthroughs in solar cell technology have led to the development of effective hybrid solar cells that have a power conversion efficiency of ~23% [101]. Furthermore, the recent discovery of dye-sensitized solar cells (DSCs) has challenged conventional solid-state photovoltaic technologies, functioning at a molecular and nano level [102]. In addition, combining organic and non-organic materials to improve the overall properties of solar cells is considered the future of solar cell research [103]. The future scope of solar cell technology includes the development of training programs focusing on occupational safety and health (OSH) for different types of solar PV projects, as contextual requirements vary [104]. Moreover, the fabrication of low-cost, perovskite solar cells with an efficiency of around 25% suggests the possibility for future breakthroughs in this field [105]. Also, the use of metal-organic frameworks (MOFs) in solar cells is an important field of research, showing the potential for breakthroughs in composite materials and nanotechnology [106]. Almost every country in the world is focusing on creating green energy through solar energy, with a few countries providing the most power through solar energy. Despite these advances, scientific obstacles remain in the ongoing development of solar cells[107–110]. Here are some current challenges and future perspectives in the field of solar cell technology:

a. Material limitations: Silicon, the most common material used in solar cells, has certain physical as well as economic constraints. High manufacturing costs are incurred due to the necessity of achieving a high level of purity and a crystalline structure for maximum performance, while the efficiency of silicon-based solar cells is subject to theoretical and practical limits [111].
b. Efficiency constraints: Solar cell efficiency has increased considerably over time, yet there is always potential for further enhancement. Mainstream silicon-based solar cells now achieve around 22%–26% efficiency for commercial use. Ongoing research is focused on developing novel materials and designs to enhance this efficiency. The Shockley-Queisser limit is the theoretical maximum efficiency achievable by single-junction solar cells, around 33.7% [112].
c. Cost impediments: The overall cost of solar cells and related technologies such as inverters and installation has significantly lowered over time. Reducing costs is a priority to enhance the competitiveness of solar energy compared to traditional fossil fuel sources [113].
d. Durability and reliability: Solar panels must endure severe environmental conditions over their lifetime of 25–30 years. Improving durability and reliability to guarantee minimum deterioration over time is a continuous challenge [114].
e. Energy storage integration: Intermittency is a concern for solar energy owing to differences in weather and daylight hours. Integrating effective energy storage options such as batteries to store surplus energy generated during peak production times is vital for enhanced grid integration and dependability [115].
f. Material availability and sustainability: Some proficient solar cell technologies depend on rare or hazardous elements, which may pose issues in terms

of resource availability, environmental effect, and recycling. Developing solar cells utilizing available, non-toxic materials is vital for sustainability [116].

g. Manufacturing scalability: Scaling up production while retaining quality and efficiency is another difficulty. Developing manufacturing technologies that can manufacture solar cells in large quantities quickly and cost-effectively is vital for broad adoption [117].
h. Space efficiency: Traditional solar panels need substantial space, which limits their deployment in urban areas or on buildings with limited roof space. Developing more space-efficient solar technology or integrating solar cells into building materials might solve this difficulty [118].
i. Incident angle sensitivity: One of the major problems with solar panels is that they lose their efficiency when sunlight strikes them at an angle. Addressing this sensitivity to incidence angles might enhance total energy collection, particularly in places with less direct sunlight [119].
j. Flexible and transparent solar cells: Developing solar cells that are flexible, lightweight, and transparent brings up new opportunities for integration into different surfaces such as windows, automobiles, and clothes. However, maintaining efficiency and durability in such systems is challenging [120,121].
k. Regulatory and policy issues: Regulatory frameworks, incentives, and policies are crucial factors in the acceptance of solar energy. To promote the extensive use of solar technology, it is essential to overcome bureaucratic obstacles, simplify regulatory procedures, and provide unwavering governmental backing [122].

The expansion of the solar energy sector is propelled by continuous advancements in materials science, manufacturing processes, and system integration, which are further addressed through research and development.

6.6 FUTURE SCOPE IN SOLAR CELL

a. Breakthrough in efficiency: Continually, research efforts are devoted to the development of novel structures and materials that improve the performance of solar cells. For instance, perovskite solar cells have demonstrated considerable potential in attaining increased efficiencies and decreased manufacturing expenses [123,124]. The utilization of quantum dots and multi-junction cells, which employ stacked layers to absorb distinct regions of the solar spectrum, are examples of cutting-edge ideas that are aimed at exceeding the Shockley-Queisser limit. In addition to improving efficiency, concentrator photovoltaics (CPVs) do so by concentrating sunlight onto a more compact and efficient solar cell.
b. Cost reduction strategies: Costs are anticipated to decrease further due to advancements in manufacturing processes, economies of scale, and the creation of cheaper materials. Moreover, solar energy is becoming more accessible through the implementation of novel business models and financing strategies [125].

c. Thin-film solar cells: By using thin layers of photovoltaic materials, thin-film solar cells provide benefits such as lightweight and flexibility and also reduce manufacturing costs. Further research and advancement in this field may result in solar panels making them more cost-effective and efficient [126].
d. Bifacial solar panels: Bifacial solar panels have the capability to utilize the reflected light in order to enhance energy generation by capturing sunlight from both the front and rear surfaces. Potential advancements in bifacial technology and modeling may increase the efficiency and cost-effectiveness of these panels [127].
e. Energy storage solutions: Advances in energy storage technologies, such as flow batteries, lithium-ion batteries, and hydrogen storage, will play an important role in facilitating increased integration of solar energy into the grid [128].
f. Perovskite-silicon tandem cells: Combining the perovskite and silicon solar cells into tandem structures could maximize the advantages of both materials, delivering better efficiencies while maintaining stability and scalability [129,130].
g. Solar PV integrated systems: Integrating solar panels into building materials, such as roofing shingles, windows, and facades, could expand the adoption of solar energy by reducing installation costs and enhancing aesthetics [131].
h. Artificial intelligence and machine learning: Artificial intelligence and machine learning algorithms are being used to optimize the design, manufacturing, operation, and maintenance of solar energy systems, leading to improved efficiency and reliability [132].

6.7 CONCLUSIONS

Solar energy has made significant strides in recent years, but there are still challenges to overcome and exciting opportunities for further innovation and advancement. Continued research, investment, and collaboration across academia, industry, and government will be essential to realizing the full potential of solar cell technology. In conclusion, the investigation of the growth of the solar cell industry and its varied uses highlights the critical part that solar energy plays in our journey to a sustainable future. This chapter has given readers insightful knowledge of the difficulties, marketplace, and cutting-edge uses that are reshaping the solar sector.

First, we highlighted the main obstacles on the national and global stages, such as ambiguous regulations and technology limitations. In order to establish an environment that is favorable for the adoption of solar energy, governments, industry stakeholders, and researchers must work together to overcome these obstacles. The market for solar cells has experienced exponential development due to falling costs and rising efficiency. This expansion establishes solar technology as a pillar of our energy landscape and signals a dramatic shift towards renewable energy sources. The wide range of uses for solar cells that are presented in this chapter demonstrate the adaptability and promise of solar energy in a number of industries. Solar technology is transforming the way we create and use electricity, from powering residential rooftops to allowing space-based power plants and transparent solar cells integrated into architectural designs. The potential applications of solar cell technology seem bright. Further developments in the fields of materials research, manufacturing, and energy storage will raise the efficiency, dependability, and cost-effectiveness of solar cells. Solar energy will have a significant beneficial influence on reducing climate change and fostering energy independence as it becomes more widely available and incorporated into our everyday lives.

In summary, the chapter highlights the revolutionary nature of solar cell technologies and the necessity of resolving present issues in order to realize their full potential. Solar energy will be essential to creating a better, more sustainable energy future for future generations via persistent innovation and group efforts.

NOTE

* Corresponding author: Summit Jalota, Department of Physics, D.A.V. College, Jalandhar, Punjab, India.

REFERENCES

1. P. Choudhary, & R. K. Srivastava. Sustainability perspectives-A review for solar photovoltaic trends and growth opportunities. *Journal of Cleaner Production* 227 (2019): 589–612.
2. M. Gul, Y. Kotak, & T. Muneer. Review on recent trend of solar photovoltaic technology. *Energy Exploration & Exploitation* 34, no. 4 (2016): 485–526.
3. A. Machín, & F. Márquez. Advancements in photovoltaic cell materials: Silicon, organic, and perovskite solar cells. *Materials* 17, no. 5 (2024): 1165.
4. M. Shen, W. Huang, M. Chen, B. Song, G. Zeng, & Y. Zhang. (Micro) plastic crisis: Un-ignorable contribution to global greenhouse gas emissions and climate change. *Journal of Cleaner Production* 254 (2020): 120138.
5. U. F. Akpan, & G. E. Akpan. The contribution of energy consumption to climate change: A feasible policy direction. *International Journal of Energy Economics and Policy* 2, no. 1 (2012): 21–33.
6. M. O'g'li, & X. Faxri. Renewable energy sources: Advancements, challenges, and prospects. *International Journal of Advance Scientific Research* 3(08) (2023): 14–25.
7. J. Gao, S. Kovats, S. Vardoulakis, P. Wilkinson, A. Woodward, J. Li, S. Gu, et al. Public health co-benefits of greenhouse gas emissions reduction: A systematic review. *Science of the Total Environment* 627 (2018): 388–402.
8. N. Kant, & P. Singh. Review of next generation photovoltaic solar cell technology and comparative materialistic development. *Materials Today: Proceedings*, 56 (2022): 3460–3470.
9. J. Szlufcik, S. Sivoththaman, J. F. Nlis, R. P. Mertens, & R. Van Overstraeten. Low-cost industrial technologies of crystalline silicon solar cells. *Proceedings of the IEEE* 85, no. 5 (1997): 711–730.
10. I. M. Alarifi. Advanced selection materials in solar cell efficiency and their properties-A comprehensive review. *Materials Today: Proceedings* 81 (2023): 403–414.
11. E. K. Solak, & E. Irmak. Advances in organic photovoltaic cells: A comprehensive review of materials, technologies, and performance. *RSC Advances* 13, no. 18 (2023): 12244–12269.

12. X. Ju, C. Xu, Y. Hu, X. Han, G. Wei, & X. Du. A review on the development of photovoltaic/concentrated solar power (PV-CSP) hybrid systems. *Solar Energy Materials and Solar Cells* 161 (2017): 305–327.
13. Y. Cui, Z. Geng, Q. Zhu, & Y. Han. Multi-objective optimization methods and application in energy saving. *Energy* 125 (2017): 681–704.
14. Z. Liu, X. Zhang, Y. Sun, & Y. Zhou. Advanced controls on energy reliability, flexibility, resilience, and occupant-centric control for smart and energy-efficient buildings—A state-of-the-art review. *Energy and Buildings* 297 (2023): 113436.
15. S. Pandey, V. S. Singh, N. P. Gangwar, M. M. Vijayvergia, C. Prakash, & D. N. Pandey. Determinants of success for promoting solar energy in Rajasthan, India. *Renewable and Sustainable Energy Reviews* 16, no. 6 (2012): 3593–3598.
16. N. K. Sharma, P. K. Tiwari, & Y. R. Sood. Solar energy in India: Strategies, policies, perspectives and future potential. *Renewable and Sustainable Energy Reviews* 16, no. 1 (2012): 933–941.
17. A. Kumar, K. Kumar, N. Kaushik, S. Sharma, & S. Mishra. Renewable energy in India: Current status and future potentials. *Renewable and Sustainable Energy Reviews* 14, no. 8 (2010): 2434–2442.
18. S. Mittal, W. Ahmed, & C. S. Koli. Renewable energy in India: Current status and future potentials. *Invertis Journal of Renewable Energy* 8, no. 2 (2018): 96–108.
19. M. Goel. Solar rooftop in India: Policies, challenges and outlook. *Green Energy & Environment* 1, no. 2 (2016): 129–137.
20. T. Muneer, M. Asif, & S. Munawwar. Sustainable production of solar electricity with particular reference to the Indian economy. *Renewable and Sustainable Energy Reviews* 9, no. 5 (2005): 444–473.
21. G. A. Heath, T. J. Silverman, M. Kempe, M. Deceglie, D. Ravikumar, T. Remo, H. Cui, et al. Research and development priorities for silicon photovoltaic module recycling to support a circular economy. *Nature Energy* 5, no. 7 (2020): 502–510.
22. D. Sarti, & R. Einhaus. Silicon feedstock for the multi-crystalline photovoltaic industry. *Solar Energy Materials and Solar Cells* 72, no. 1–4 (2002): 27–40.
23. S. Bernardis. Engineering impurity behavior on the micron-scale in metallurgical-grade silicon production. PhD diss., Massachusetts Institute of Technology, 2012.
24. J. Safarian. Thermochemical aspects of boron and phosphorus distribution between silicon and BaO-SiO_2 and CaO-BaO-SiO_2 lags. *Silicon* 11 (2019): 437–451.
25. Z. Zhang, S. Zhai, D. Wei, X. Guo, H. Li, Y. Zhuang, & P. Xing. Enhancing the separation of Fe/Al/Ca impurities from metallurgical silicon to purify by slagging-directional solidification. *Vacuum* 215 (2023): 112361.
26. C. del Cañizo Nadal, S. Binetti, & T. Buonassisi. 1 Purity Requirements for Silicon in Photovoltaic Applications. In B. Ceccaroli, E. Ovrelid, S. Pizzini (eds.), *Solar Silicon Processes: Technologies, Challenges, and Opportunities*, pp. 1–47. CRC Press, 2016.
27. Z.-J. Bao, H.-Y. Yang, B.-X. Dong, F. Chang, C.-D. Li, Y. Jiang, L.-Y. Chen, et al. Development trend in composition optimization, microstructure manipulation, and strengthening methods of die steels under lightweight and integrated die casting. *Materials* 16, no. 18 (2023): 6235.
28. H. J. Park, J. Kim, D. Choi, S.-W. Lee, D. Kang, H.-S. Lee, D. Kim, et al. Boron-doped polysilicon using spin-on doping for high-efficiency both-side passivating contact silicon solar cells. *Progress in Photovoltaics: Research and Applications* 31, no. 5 (2023): 461–473.
29. Y. Yu, S. Li, F. Xi, H. Lan, D. Wu, Z. Chen, W. Ma, et al. Influence of the structural differences between end-of-life Al-BSF and PERC modules on the Al leaching separation behavior. *Solar Energy* 263 (2023): 111938.
30. Y. Zhao, P. Procel, C. Han, L. Cao, G. Yang, E. Özkol, A. Alcañiz, et al. Strategies for realizing high-efficiency silicon heterojunction solar cells. *Solar Energy Materials and Solar Cells* 258 (2023): 112413.

31. P. M. Tembo, & V. Subramanian. Current trends in silicon-based photovoltaic recycling: A technology, assessment, and policy review. *Solar Energy* 259 (2023): 137–150.
32. M. Ramzan, S. A. Raza, M. Usman, G. D. Sharma, & H. A. Iqbal. Environmental cost of non-renewable energy and economic progress: Do ICT and financial development mitigate some burden? *Journal of Cleaner Production* 333 (2022): 130066.
33. C. Ballif, F.-J. Haug, M. Boccard, P. J. Verlinden, & G. Hahn. Status and perspectives of crystalline silicon photovoltaics in research and industry. *Nature Reviews Materials* 7, no. 8 (2022): 597–616.
34. C. A. Wolden, J. Kurtin, J. B. Baxter, I. Repins, S. E. Shaheen, J. T. Torvik, A. A. Rockett, et al. Photovoltaic manufacturing: Present status, future prospects, and research needs. *Journal of Vacuum Science & Technology A* 29, no. 3 (2011): 030801.
35. V. Stratiotou Efstratiadis, & N. Michailidis. Sustainable recovery, recycle of critical metals and rare earth elements from waste electric and electronic equipment (circuits, solar, wind) and their reusability in additive manufacturing applications: A review. *Metals* 12, no. 5 (2022): 794.
36. T. M. Gür. Review of electrical energy storage technologies, materials and systems: Challenges and prospects for large-scale grid storage. *Energy & Environmental Science* 11, no. 10 (2018): 2696–2767.
37. A. Codemo, A. Barbini, A. Mantouza, A. Bitziadis, & R. Albatici. Integration of public perception in the assessment of licensed solar farms: A case study in Greece. *Sustainability* 15, no. 13 (2023): 9899.
38. D. Hart, & K. Birson. Deployment of solar photovoltaic generation capacity in the United States. *Office of Energy Policy and Systems Analysis U.S. Department of Energy* (2016).
39. A. R. K. Gollakota, & C.-M. Shu. COVID-19 and energy sector: Unique opportunity for switching to clean energy. *Gondwana Research* 114 (2023): 93–116.
40. G. L. Barbose, & N. R. Darghouth. Tracking the sun: Pricing and design trends for distributed photovoltaic systems in the United States-2019 edition, LBL Publications, 2019. https://escholarship.org/uc/item/5422n7wm
41. J. Ball, D. Reicher, X. Sun, & C. Pollock. The new solar system: China's evolving solar industry and its implications for competitive solar power in the United States and the world. Stanford Univ., CA (United States), 2017.
42. V. Bandi, T. Sahrakorpi, J. Paatero, & R. Lahdelma. The paradox of mini-grid business models: A conflict between business viability and customer affordability in rural India. *Energy Research & Social Science* 89 (2022): 102535.
43. M. Schneiberg, & T. Bartley. Organizations, regulation, and economic behavior: Regulatory dynamics and forms from the nineteenth to twenty-first century. *Annual Review of Law and Social Science* 4 (2008): 31–61.
44. P. Choudhary, & Srivastava, R. K. Sustainability perspectives-A review for solar photovoltaic trends and growth opportunities. *Journal of Cleaner Production* 227 (2019): 589–612.
45. A. Yilanci, I. Dincer, & H. K. Ozturk. A review on solar-hydrogen/fuel cell hybrid energy systems for stationary applications. *Progress in Energy and Combustion Science* 35, no. 3 (2009): 231–244.
46. I. Dincer, & C. Zamfirescu. Sustainable hydrogen production options and the role of IAHE. *International Journal of Hydrogen Energy* 37, no. 21 (2012): 16266–16286.
47. A. Franco, & C. Giovannini. Recent and future advances in water electrolysis for green hydrogen generation: Critical analysis and perspectives. *Sustainability* 15, no. 24 (2023): 16917.
48. H. Matsumoto. Space solar power satellite/station and the politics. *IEICE Proceedings Series* 14, no. 21Q1-4 (2009): 37–39.
49. R. Venugopal, H. R. Manjunath, & A. V. Raghu. Overview of space based solar power. *Material Science Research India* 19, no. 2 (2022). http://dx.doi.org/10.13005/msri/190203

50. M. M. Olaimat, V. Nayyeri, & O. M. Ramahi. Simulation and optimization of rectenna systems. *Scientific Reports* 13, no. 1 (2023): 17107.
51. S. Sasaki, K. Tanaka, & K.-I. Maki. Microwave power transmission technologies for solar power satellites. *Proceedings of the IEEE* 101, no. 6 (2013): 1438–1447.
52. E. A. Abdelaziz, R. Saidur, & S. Mekhilef. A review on energy saving strategies in industrial sector. *Renewable and Sustainable Energy Reviews* 15, no. 1 (2011): 150–168.
53. A. M. Abdel-Dayem, & M. A. Mohamad. Potential of solar energy utilization in the textile industry—A case study. *Renewable Energy* 23, no. 3–4 (2001): 685–694.
54. M. I. Ismail, N. A. Yunus, & H. Hashim. Integration of solar heating systems for low-temperature heat demand in food processing industry–A review. *Renewable and Sustainable Energy Reviews* 147 (2021): 111192.
55. S. A. Kalogirou. Solar thermal collectors and applications. *Progress in Energy and Combustion Science* 30 (2004): 231–295.
56. G. G. Maidment, & A. Paurine. Solar Cooling and Refrigeration Systems. In A. A. M. Sayigh (ed.), *Comprehensive Renewable Energy*, pp. 481–494. Elsevier, 2012.
57. T. Kodama. High-temperature solar chemistry for converting solar heat to chemical fuels. *Progress in Energy and Combustion Science* 29 (2003): 567.
58. I. Dincer, & T. A. H. Ratlamwala. Solar Thermal Power Systems. In S. A. Elias (ed.), *Reference Module in Earth Systems and Environmental Sciences*, Chapter 05931. Elsevier, 2013.
59. A. Reco, E. Gundabattini, D. S. Gnanaraj, & C. Masselli. A comparative study on the performances of flat plate and evacuated tube collectors deployable in domestic solar water heating systems in different climate areas. *Climate* 8 (2020): 78.
60. G. N. Tiwari, Y. P. Yadav, P. C. Eames, & B. Nortan. Solar distillation systems: The state of the art in design development and performance analysis. *Renewable Energy* 5 (1994): 509–516.
61. V. Manikandan, K. Shanmugasundaram, S. Shanmugam, B. Janarthanana, & J. Chandrasekaran. Wick type solar stills: A review. *Renewable and Sustainable Energy Reviews* 20 (2013): 322–335.
62. H. Sharon, & K. S. Reddy. A review of solar energy driven desalination technologies. *Renewable and Sustainable Energy Reviews* 41 (2015): 1080–1118.
63. K. R. Kumar, N. V. V. K. Chaitanya, & N. S. Kumar. Solar thermal energy technologies and its applications for process heating and power generation – A review. *Journal of Cleaner Production* 282 (2021): 125296.
64. S. Kalogirou. The potential of solar industrial process heat applications. *Applied Energy* 76 (2003): 337–361.
65. S. Pati, & O. P. Verma. Energy integration of solar assisted multiple stage evaporator and optimum parameter selection. *Energy* 239 (2022): 122162.
66. S. Indora, & T. C. Kandpal. Institutional cooking with solar energy: A review. *Renewable and Sustainable Energy Reviews* 84 (2018): 131–154.
67. V. P. Chandramohan, & P. Das. Low and Medium Temperature Solar Thermal Collectors. In T. V. Arjunan, V. Selvaraj, M. M. Matheswaran (eds.), *Solar Thermal Conversion Technologies for Industrial Process Heating*, pp. 37–74. CRC Press, 2023.
68. P. P. Otte. A (new) cultural turn toward solar cooking—Evidence from six case studies across India and Burkina Faso. *Energy Research & Social Science* 2 (2014): 49–58.
69. T. Kanyowa, G. Victor Nyakujara, E. Ndala, & S. Das. Performance analysis of Scheffler dish type solar thermal cooking system cooking 6000 meals per day. *Solar Energy* 218 (2021): 563–570.
70. H. Naik, P. Baredar, & A. Kumar. Medium temperature application of concentrated solar thermal technology: Indian perspective. *Renewable and Sustainable Energy Reviews* 76 (2017): 369–378.

71. International Energy Agency. Energy Access Outlook 2017: From poverty to prosperity. 2017.
72. International Energy Agency. Trends 2015 in photovoltaic applications. 2015.
73. M. Gallucci. Energy equity: Bringing solar power to low-income communities. *Yale Environment* 360 (2019): 2019.
74. T. Urmee, & D. Harries. A survey of solar PV program implementers in Asia and the Pacific regions. *Energy for Sustainable Development* 13 (2009): 24–32.
75. S. Karekezi, & W. Kithyoma. Renewable energy strategies for rural Africa: Is a PV-led renewable energy strategy the right approach for providing modern energy to the rural poor to sub-Saharan Africa? *Energy Policy* 30 (2002): 1071–1086.
76. https://www.solarreviews.com/blog/lead-acid-batteries-for-solar-storage
77. S. M. Jawd, N. H. Chekhyor, & A. Mahmood. Types of solar cell batteries and their energy charging methods. *Journal of Thermal Engineering and Applications* 8 (2021): 16–22.
78. C. Udayasri, M. Lavanya, & M. Arivalagan. Solar based adjustable voltage regulator using IC LM317. *Journal of Chemical and Pharmaceutical Sciences* (2016): 273–275. ISSN: 0974–2115
79. B. Sigrin, & M. Mooney. *Rooftop Solar Technical Potential for Low-to-Moderate Income Households in the United States.* Golden, CO: National Renewable Energy Laboratory. NREL/TP-6A20–70901, 2018.
80. Liu, W., Liu, Y., Yang, Z., Xu, C., Li, X., Huang, S., Shi, J., et al. Flexible solar cells based on foldable silicon wafers with blunted edges. *Nature* 617 (2023): 717–723.
81. https://www.motortrend.com/features/the-2023-toyota-prius-primes-battery-could-take-three-weeks-to-recharge/
82. X. Li, P. Li, Z. Wu, D. Luo, H. Y. Yu, & Z. H. Lu. Review and perspective of materials for flexible solar cells. *Materials Reports: Energy* 1 (2021): 100001.
83. E. W. Williams, K. Jones, A. J. Griffiths, D. J. Roughley, J. M. Bell, J. H. Steven, M. J. Huson, et al. The electrophoresis of thin Film CdS/Cu_2S solar cells. *Solar Cells* 1, no. 4 (1980): 357–366.
84. Y. Xiao, G. Han, H. Zhou, & J. Wu. An efficient titanium foil based perovskite solar cell: Using a titanium dioxide nanowire array anode and transparent poly(3,4-ethylenedioxythiophene) electrode. *RSC Advances* 6 (2016): 2778–2784.
85. M. Kim, H.-G. Yun, L.-W. Jang, D.-W. Jeon, M. G. Kang, J.-H. Yoon, J.-M. Kim, et al. Promising efficiency enhancement in cobalt redox couple-based back-illuminated dye-sensitized solar cells with titanium foil substrate. *Journal of Power Sources* 278 (2015): 32–37.
86. S. Sheehan, P. K. Surolia, O. Byrne, S. Garner, P. Cimo, X. Li, D. P. Dowling, et al. Flexible glass substrate based dye sensitized solar cells. *Solar Energy Materials and Solar Cells* 132 (2015): 237–244.
87. C.-Y. Peng, T. P. Dhakal, S. Garner, P. Cimo, S. Lu, & C. R. Westgate. Fabrication of Cu_2ZnSnS_4 solar cell on a flexible glass substrate. *Thin Solid Films* 562 (2014): 574–577.
88. H. C. Weerasinghe, F. Huang, & Y.-B. Cheng. Fabrication of flexible dye sensitized solar cells on plastic substrates. *Nano Energy* 2 (2013): 174–189.
89. Q. Lin, L. Lu, M. M. Tavakoli, C. Zhang, G. C. Lui, Z. Chen, X. Chen, et al. High performance thin film solar cells on plastic substrates with nanostructure-enhanced flexibility. *Nano Energy* 22 (2016): 539–547.
90. R. Dallaev, T. Pisarenko, N. Papež, & V. Holcman. Overview of the current state of flexible solar panels and photovoltaic materials. *Materials* 16 (2023): 5839.
91. R. R. Mutagekar, S. A. Kachare, P. S. Babre, G. D. Bhirud, H. N. Pendor, & A. V. Shinde. A review on transparent solar panels. Preprints 2023, 2023101865.
92. V. R. Mehta, & N. M. Ravindra. Screen printing to 3D printing of solar cells—An overview. In Z. Peng, J.-Y. Hwang, J. Downey, D. Gregurek, B. Zhao, O. Yucel, E. Keskinkilic, T. Jiang, J. White, M. Mahmoud (eds.), *TMS 2020 149th Annual Meeting & Exhibition Supplemental Proceedings*, pp. 1935–1944. Springer International Publishing, 2020.

93. S. Ito. Printable solar cells. *Wiley Interdisciplinary Reviews: Energy and Environment* 4 (2015): 51–73.
94. S. Ganesan, S. Mehta, & D. Gupta. Fully printed organic solar cells-A review of techniques, challenges and their solutions. *Opto-Electronics Review* 27 (2019): 298–320.
95. R. R. Lunt, & V. Bulovic. Transparent, near-infrared organic photovoltaic solar cells for window and energy-scavenging applications. *Applied Physics Letters* 98 (2011): 113305.
96. A. A. F. Husain, W. Z. W. Hasan, S. Shafie, M. N. Hamidon, & S. S. Pandey. A review of transparent solar photovoltaic technologies. *Renewable and Sustainable Energy Reviews* 94 (2018): 779–791.
97. R. Betancur, P. Romero-Gomez, A. Martinez-Otero, X. Elias, M. Maymó, & J. Martorell. Transparent polymer solar cells employing a layered light-trapping architecture. *Nature Photonics* 7, no. 12 (2013): 995–1000.
98. K.-S. Chen, J.-F. Salinas, H.-L. Yip, L. Huo, J. Hou, & A. K-Y. Jen. Semi-transparent polymer solar cells with 6% PCE, 25% average visible transmittance and a color rendering index close to 100 for power generating window applications. *Energy & Environmental Science* 5, no. 11 (2012): 9551–9557.
99. C. Roldan-Carmona, O. Malinkiewicz, R. Betancur, G. Longo, C. Momblona, F. Jaramillo, L. Camacho, et al. High efficiency single-junction semitransparent perovskite solar cells. *Energy & Environmental Science* 7, no. 9 (2014): 2968–2973.
100. https://www.anker.com/blogs/solar/solar-powered-calculator
101. A. M. Elseman. Organometal Halide Perovskites Thin Film and Their Impact on the Efficiency of Perovskite Solar Cells. In A. G. Avila Bernal, J. A. Perez-Taborda (eds.), *Coatings and Thin-Film Technologies*, pp. 223–240. Intech Open, 2018.
102. P. D. Sekaran, & R. Marimuthu. An extensive analysis of dye-sensitized solar cell (DSSC). *Brazilian Journal of Physics* 54, no. 1 (2024): 28.
103. Z. Shao. Comparison and analysis of the third generation of the solar cells. *Highlights in Science, Engineering and Technology* 27 (2022): 498–505.
104. A. Sen, A. S. Mohankar, A. Khamaj, & S. Karmakar. Emerging OSH issues in installation and maintenance of floating solar photovoltaic projects and their link with sustainable development goals. *Risk Management and Healthcare Policy* (2021): 1939–1957.
105. M. I. Hossain, A. M. Saleque, S. Ahmed, I. Saidjafarzoda, M. Shahiduzzaman, W. Qarony, D. Knipp, et al. Perovskite/perovskite planar tandem solar cells: A comprehensive guideline for reaching energy conversion efficiency beyond 30%. *Nano Energy* 79 (2021): 105400.
106. J. Dou, & Q. Chen. MOFs in emerging solar cells. *Chinese Journal of Chemistry* 41, no. 6 (2023): 695–709.
107. M.-E. Ragoussi, & T. Torres. New generation solar cells: Concepts, trends and perspectives. *Chemical Communications* 51, no. 19 (2015): 3957–3972.
108. M. F. L. Abdullah, & A. Z. M. Rus. Dye sensitized solar cell (DSSC) performance analysis based on series circuit. *Applied Mechanics and Materials* 773 (2015): 691–695.
109. L. Wang, M. Al-Mamun, P. Liu, Y. Wang, H. G. Yang, H. F. Wang, & H. Zhao. The search for efficient electrocatalysts as counter electrode materials for dye-sensitized solar cells: Mechanistic study, material screening and experimental validation. *Nature Publishing Group Asia Materials* 7, no. 11 (2015): e226–e226.
110. H. Zhang, C. Liang, Y. Zhao, M. Sun, H. Liu, J. Liang, D. Li, et al. Dynamic interface charge governing the current–voltage hysteresis in perovskite solar cells. *Physical Chemistry Chemical Physics* 17, no. 15 (2015): 9613–9618.
111. M. A. Green, K. Emery, Y. Hishikawa, W. Warta, E. D. Dunlop, D. H. Levi, & A. W. Y. Ho-Baillie. Solar cell efficiency tables (version 49). *Progress in Photovoltaics: Research and Applications* 25, no. 1 (2017): 3–13.

112. W. Shockley, & H. J. Queisser. Detailed balance limit of efficiency of p-n junction solar cells. *Journal of Applied Physics* 32, no. 3 (1961): 510–519.
113. R. Fu, D. Feldman, & R. Margolis. U.S. Solar Photovoltaic System Cost Benchmark: Q1 2017. National Renewable Energy Laboratory (NREL) Report No. NREL/TP-6A20-68925, 2017.
114. L. M. Shaker, A. Alamiery, W. N. R. Wan Isahak, & W. Al-Azzawi. Corrosion in solar cells: Challenges and solutions for enhanced performance and durability. *Journal of Optics* (2023): 1–15.
115. P. Denholm, E. G. Ela, B. Kirby, & M. Milligan. The Role of Energy Storage with Renewable Electricity Generation. National Renewable Energy Laboratory (NREL), Technical Report NREL/TP-6A2-47187, 2010.
116. Wang, Li, Y. Zhang, M. Kim, M. Wright, R. Underwood, R. S. Bonilla, & B. Hallam. Sustainability evaluations on material consumption for terawatt-scale manufacturing of silicon-based tandem solar cells. *Progress in Photovoltaics: Research and Applications* 31 (2023): 1442–1454.
117. L. Klein, S. Ziegler, F. Laufer, C. Debus, M. Götz, K. Maier-Hein, U. W. Paetzold, et al. Discovering process dynamics for scalable perovskite solar cell manufacturing with explainable AI. *Advanced Materials* 36 (2023): 2307160.
118. A. Sharma, & V. K. Bajpai. A review on development and technology of various types of solar PV cell. In D. Panchal, P. Chatterjee, M. Tyagi, R. P. Singh (eds.), *Optimization Methods for Engineering Problems*, p. 215. Apple Academic Press Inc, 2023.
119. M. K. Sharma, & J. Bhattacharya. Dependence of spectral factor on angle of incidence for monocrystalline silicon based photovoltaic solar panel. *Renewable Energy* 184 (2022): 820–829.
120. L. Huang, Y. Li, Z. Zheng, Y. Bai, T. P. Russell, & C. He. Flexible, transparent, and sustainable cellulose-based films for organic solar cell substrates. *Materials Horizons* 11 (2024): 1560–1566.
121. R. Tian, S. Zhou, Y. Meng, C. Liu, & Z. Ge. Material and device design of flexible perovskite solar cells for next-generation power supplies. *Advanced Materials* 36 (2024): 2311473.
122. A. Sharma, P. Mahajan, & R. Garg. End-of-life solar photovoltaic panel waste management in India: Forecasting and environmental impact assessment. *International Journal of Environmental Science and Technology* 21, no. 2 (2024): 1961–1980.
123. R. R. King, D. C. Law, K. M. Edmondson, C. M. Fetzer, G. S. Kinsey, H. Yoon, R. A. Sherif, et al. 40% efficient metamorphic GaInP/GaInAs/Ge multijunction solar cells. *Applied Physics Letters* 90, no. 18 (2007): 183516.
124. T. Buchmann, & P. Wolf. Breakthrough inventions in solar PV and wind technologies: The role of scientific discoveries. *Technological Forecasting and Social Change* 198 (2024): 122975.
125. B. S. Kumar, B. M. Kunar, & C. S. N. Murthy. An adaptive modeling for bifacial solar module levelized cost and performance analysis for mining application. *Progress in Photovoltaics: Research and Applications* 32, no. 3 (2024): 186–198.
126. A. Maalouf, T. Okoroafor, Z. Jehl, V. Babu, & S. Resalati. A comprehensive review on life cycle assessment of commercial and emerging thin-film solar cell systems. *Renewable and Sustainable Energy Reviews* 186 (2023): 113652.
127. M. Vimala, G. Ramadas, M. Perarasi, A. M. Manokar, & R. Sathyamurthy. A review of different types of solar cell materials employed in bifacial solar photovoltaic panel. *Energies* 16, no. 8 (2023): 3605.
128. M. A. Rahman, J.-H. Kim, & S. Hossain. Recent advances of energy storage technologies for grid: A comprehensive review. *Energy Storage* 4, no. 6 (2022): e322.

129. E. Aydin, E. Ugur, B. K. Yildirim, T. G. Allen, P. Dally, A. Razzaq, F. Cao, et al. Enhanced optoelectronic coupling for perovskite/silicon tandem solar cells. *Nature* 623, no. 7988 (2023): 732–738.
130. S. Mariotti, E. Köhnen, F. Scheler, K. Sveinbjörnsson, L. Zimmermann, M. Piot, F. Yang, et al. Interface engineering for high-performance, triple-halide perovskite–silicon tandem solar cells. *Science* 381, no. 6653 (2023): 63–69.
131. A. S. R. Bati, Y. L. Zhong, P. L. Burn, M. K. Nazeeruddin, P. E. Shaw, & M. Batmunkh. Next-generation applications for integrated perovskite solar cells. *Communications Materials* 4, no. 1 (2023): 2.
132. D. Mhlanga. Artificial intelligence and machine learning for energy consumption and production in emerging markets: A review. *Energies* 16, no. 2 (2023): 745.

Index

Note: **Bold** page numbers refer to tables and *italic* page numbers refer to figures.

For Product Safety Concerns and Information please contact our EU
representative GPSR@taylorandfrancis.com
Taylor & Francis Verlag GmbH, Kaufingerstraße 24, 80331 München, Germany

www.ingramcontent.com/pod-product-compliance
Ingram Content Group UK Ltd.
Pitfield, Milton Keynes, MK11 3LW, UK
UKHW022123190926
475702UK00013B/1621
9781032814087